POWERED PROSTHESES

POWERED PROSTHESES

Design, Control, and Clinical Applications

Edited by

HOUMAN DALLALI

EMEL DEMIRCAN

MO RASTGAAR

ACADEMIC PRESS

An imprint of Elsevier

Academic Press is an imprint of Elsevier
125 London Wall, London EC2Y 5AS, United Kingdom
525 B Street, Suite 1650, San Diego, CA 92101, United States
50 Hampshire Street, 5th Floor, Cambridge, MA 02139, United States
The Boulevard, Langford Lane, Kidlington, Oxford OX5 1GB, United Kingdom

Notices
Knowledge and best practice in this field are constantly changing. As new research and experience
broaden our understanding, changes in research methods, professional practices, or medical treatment may become necessary.

Practitioners and researchers must always rely on their own experience and knowledge in evaluating
and using any information, methods, compounds, or experiments described herein. In using such
information or methods they should be mindful of their own safety and the safety of others, including parties for whom they have a professional responsibility.

To the fullest extent of the law, neither the Publisher nor the authors, contributors, or editors, assume any liability for any injury and/or damage to persons or property as a matter of products liability, negligence or otherwise, or from any use or operation of any methods, products, instructions, or
ideas contained in the material herein.

Library of Congress Cataloging-in-Publication Data
A catalog record for this book is available from the Library of Congress

British Library Cataloguing-in-Publication Data
A catalogue record for this book is available from the British Library

ISBN: 978-0-12-817450-0

For information on all Academic Press publications
visit our website at https://www.elsevier.com/books-and-journals

Publisher: Mara Conner
Acquisitions Editor: Sonnini R. Yura
Editorial Project Manager: Emma Hayes
Production Project Manager: Surya Narayanan Jayachandran
Designer: Matthew Limbert

Typeset by Thomson Digital

Working together
to grow libraries in
developing countries

www.elsevier.com • www.bookaid.org

Contents

9. Semi-active prostheses for low-power gait adaptation   **201**
Peter Gabriel Adamczyk

Contributors

Sofiane Achiche
Mechanical Engineering Department, Polytechnique Montreal, Montreal, QC, Canada

Peter Gabriel Adamczyk
University of Wisconsin-Madison, Madison WI, United States

Olivier Barron
Mechanical Engineering Department, Polytechnique Montreal, Montreal, QC, Canada

Houman Dallali
Department of Computer Science, California State University, Channel Islands, Camarillo, CA, United States

Neil Dhir
The Alan Turing Institute British Library, London

Nafiseh Ebrahimi
Mechanical Engineering Department, University of Texas at San Antonio (UTSA), One UTSA Circle, San Antonio, TX, United States

Martin Grimmer
Lauflabor Locomotion Lab, Institute of Sport Science, Centre for Cognitive Science, Technische Universität Darmstadt, Darmstadt, Germany

Sehoon Ha
Google Brain, Mountain View, CA, United States

Hsiang Hsu
Department of Mechanical Engineering, Georgia Institute of Technology, Atlanta, GA, United States

Amir Jafari
Mechanical Engineering Department, University of Texas at San Antonio (UTSA), One UTSA Circle, San Antonio, TX, United States

Lauren N. Knop
Polytechnic Institute, Purdue University, West Lafayette, IN, United States

Hyunglae Lee
School for Engineering of Matter, Transport and Energy, Arizona State University, Tempe, AZ, United States

C. Karen Liu
School of Interactive Computing, Georgia Institute of Technology, Atlanta, GA, United States

Andrew Luo
Mechanical Engineering Department, University of Texas at San Antonio (UTSA), One
UTSA Circle, San Antonio, TX, United States

Alireza Mohammadi
Department of Electrical and Computer Engineering, University of Michigan, Dearborn,
MI, United States

Gautham Muthukumaran
Mechanical Engineering Department, University of Texas at San Antonio (UTSA), One
UTSA Circle, San Antonio, TX, United States

Amirreza Naseri
Mechanical Engineering School, Tarbiat Modares University of Tehran, Tehran, Iran

Maxime Raison
Mechanical Engineering Department, Polytechnique Montreal, Montreal, QC, Canada

Mo Rastgaar
Polytechnic Institute, Purdue University, West Lafayette, IN, United States

Guilherme A. Ribeiro
Polytechnic Institute, Purdue University, West Lafayette, IN, United States

André Seyfarth
Lauflabor Locomotion Lab, Institute of Sport Science, Centre for Cognitive Science,
Technische Universität Darmstadt, Darmstadt, Germany

Maziar Sharbafi
Lauflabor Locomotion Lab, Institute of Sport Science, Centre for Cognitive Science,
Technische Universität Darmstadt, Darmstadt, Germany

Yun Seong Song
Department of Mechanical and Aerospace Engineering, Missouri University of Science
and Technology, Rolla, MO, United States

Lena H. Ting
Department of Biomedical Engineering, Georgia Institute of Technology and Emory
University, Atlanta, GA, United States; Department of Rehabilitation Medicine, Division of
Physical Therapy, Emory University, Atlanta, GA, United States

CHAPTER 1

Control of transhumeral prostheses based on electromyography pattern recognition: from amputees to deep learning

Olivier Barron, Maxime Raison, Sofiane Achiche

Mechanical Engineering Department, Polytechnique Montreal, Montreal, QC, Canada

1 Introduction

The lost of a limb always results in a loss of autonomy for amputees. To rectify this condition, a usual solution is the use of a prosthesis. These can either be passive, mainly for aesthetic and postural reasons, or active for added functionality. However, designing the next-generation prostheses that would suit most amputees is a big challenge, as every one of them has a unique amputation history. Indeed, the access to this population is usually limited and heterogeneous. Therefore, researchers, engineers, and clinicians still struggle to reach the unmet needs of amputees. Their lack of acceptance is substantial and results from the frustration experienced when using current prostheses [1].

The myoelectric prosthesis is one of the best choices as an active device and is the most studied by research groups [2]. There is no such device clinically viable on the market, especially for transhumeral amputees [3]. Research groups should concentrate their efforts on developing a control system that would be intuitive to use and robust to common fluctuations associated with electromyographic signal. To reach those objectives, researchers are looking at artificial intelligence, more precisely machine learning, and myoelectric signals from the amputees' arm to directly control the device.

This chapter aims to identify and explain the state of the research surrounding the control of transhumeral prosthesis. First, the current needs and available solutions for such devices are presented before describing the relation between a myoelectric prosthesis and electromyography (EMG)

Powered Prostheses
http://dx.doi.org/10.1016/B978-0-12-817450-0.00001-8

1

pattern recognition. The next sections then expose future challenges with the appearance of deep learning, big data, and new tools to ascertain the usability of a prosthesis. Next, a review of the latest development shows a lag between cutting edge technologies and transhumeral prosthesis' development. Finally, the discussion presents recommendations for future works in the domain in order to achieve a functional prosthesis.

2 Unmet needs for transhumeral amputees

For upper limbs, the level of amputation is based on the proximity of the site to the upper body and can take different names [4] as shown in Fig. 1.1. The higher the level of amputation, the bigger is the loss of functionality resulting in a more complex prosthesis.

Although the population of transradial and transhumeral amputees is similar [5], the number of solutions available between the two groups is not. Studies regarding transradial prostheses' design usually have a stronger momentum since due to their simplicity compared to transhumeral prostheses. It is easier to restore hand gestures when the forearm's muscles are accessible. For this reason, multiple products are available on the market for this level of amputation that covers a variety of needs. For a transhumeral amputee, only the arm's surface EMG (sEMG) can be used to recreate the gestures from the elbow, the forearm, the wrist, and the hand, which is quite more complex to achieve. Hence, no technological breakthrough has

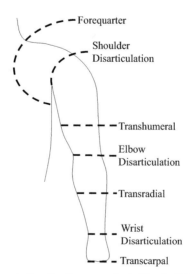

Figure 1.1 *Levels of amputation for the upper limb. (Credit: Original)*

revolutionized the field and devices available on the market are obsolete and left behind.

This situation leads to a significant lack of acceptance of the transhumeral prosthesis. About 20% of the amputees are not even trying these devices and between another 10% and 23%, end up giving up its usage [1,6]. Another study highlighted that among users that persevere, only 48% are really using their prosthesis to its full capacity [7].

Since 2005, numerous studies have investigated the lack of acceptance of transhumeral prostheses and conducted surveys among amputees. They noticed that not only their needs have been the same since the past 2 decades [4], but the developments surrounding prostheses also do not seems to take into account the lack of acceptance from amputees [8]. Moreover, these prostheses have an impact on an amputee's self-confidence and play a major role in how they behave in a social environment. It is a multidimensional problem that goes beyond simply designing a mechatronic system to serve as a tool to achieve daily tasks [9]. A study from 2016 stated that existing prostheses are indeed advanced on the mechanical and functional aspects. However, it is their unnatural [1] and unreliable control that harms the rate of adherence by amputees [10]. An unnatural control could be defined as a prosthesis' handling that requires too much physical or mental effort from its user to achieve a certain gesture. An unreliable control could be described as a prosthesis' handling that is not robust to muscle fatigue, electrode displacement, and variation in the EMG signal.

3 Available upper limb prostheses

Several types of prostheses are available, from the passive or cosmetic prosthesis to the active prosthesis able to perform complex gestures for its user. The cosmetic prosthesis has no actuator and possesses a sole purpose: to look like a human arm. It may be articulated to maintain a certain position, but the amputee would have to use his other able arm, which is impractical. With the intention of restoring most of the autonomy lost after an amputation, body-powered prostheses were the first (1955) [11] to be researched extensively. Cables are often incorporated in the prosthesis to put in motion a section of the arm when the user moves a precise part of his body (e.g., to pull your elbow backward in order to close the prosthesis' hand). However, these indirect gestures are the main reason explaining the rejection of body-powered prostheses by the amputees because they are too complex and quite unnatural to control [6].

To address this situation, myoelectric prostheses have been introduced. They are actuated with muscular electrical activity or EMG signals. As soon as the amputee wishes to move his phantom limb, her/his intent is first translated in a nerve impulse that travels naturally from the brain to the stump's targeted muscle, which then produces electrical activity. Measuring this sEMG to directly actuate a prosthesis is quite attractive since it would allow a control more intuitive and natural for the user [2]. However, despite technological breakthroughs, amputees disclose that these devices are expensive, not intuitive enough and require a lot of training to achieve sufficient control [12].

Despite this challenge, and given the popularity of transradial devices, various industries and laboratories have still commercialized their transhumeral prosthesis, such as the DynamicArm from *Ottobock* [13], Utah arm from *Motion Control* [14], and the LUKE arm from *MOBIUS bionics* [15]. However, these devices are quite expensive (>60,000 USD) and each possesses their respective deficiencies that makes them unfit to suit all transhumeral amputees. The LUKE arm, for instance, requires the stump to be very short (proximal) to fit the prosthesis whereas the Utah arm only allows the user to control one gesture at a time with sEMG. Additionally, Coapt [16] is the only company offering a clinically available socket system, the Complete Control, capable of commanding a prosthesis using sEMG [3]. Only available in the United States, it is sold without any prosthesis and rather constitutes a flexible control unit. However, as it will be presented in the next section, these studies still struggle to answer amputees' needs.

4 Myoelectric prosthesis, how does it work?

At first, myoelectric prostheses were only controlled using the amplitude of the sEMG. The command algorithm measured the signal's magnitude and activated the device according to a pre-established threshold [17]. To allow control of more than one degree of freedom at the same time, two states [18] and three states [19] command has been developed. These algorithms receive the sEMG from a single muscle and based on two or three thresholds activate different degree of freedom, accordingly. As for the prosthesis movement velocity, it can be constant whenever the device is active [20] or proportional to the mean amplitude of the sEMG [21].

Then, research groups started to use several sEMG electrodes on the stump to capture the whole electrical activity of the upper limb. The disposition of these electrodes has been debated between teams, namely placing the electrodes directly on the targeted muscles or in a matrix with equidistant electrodes [22]. Upon further investigations, results showed that when using six or more probes on the stump, no significant difference was observed with the decision algorithms with the use of either an anatomically accurate or a matrix disposition of the electrodes [2]. To record even more data about the location of those signals, high density sEMG (HD-sEMG) is being used and consists of a matrix made of numerous close-spaced electrodes that capture the electrical activity of a defined area [23].

Furthermore, to have a complete understanding of how sEMG from the stump can be used to control a prosthesis, one needs to understand how the human body copes with the loss of a limb. After an amputation, the amputee's body tries to adapt to missing or displaced nerves, to severed muscles, and to missing bones. It is pretty common for an amputee to feel like her/his limb is still present after the operation, a phenomenon known as phantom limb effect [24]. A neuromuscular rearrangement occurs after amputation and results in this sensation that tends to weaken through time but can still be felt for a lifetime.

A study from 1998 [25] revealed that amputees can experience various sensations from their phantom limb beside feeling it still attached. Some amputee experiences their phantom limb moving abruptly, like a reflex, others can experience hot and cold sensation and others can even feel pain. Besides, amputees are able to voluntarily and repeatedly move their phantom limb [26]. This sensation and the way their phantom limb moves aren't the same as if they had their whole limb. Moreover, this phantom gesture produces sEMG that has no similarities with the signal normally observed in a healthy subject [27]. Each amputee possesses her/his own unique activation pattern that depends on how the wound was closed and how the neuromuscular rearrangement occurred. Another study [28] has shown that these patterns are reproducible and reliable between sessions of sEMG acquisitions.

These patterns are the basic principle behind current myoelectric prosthesis [29]. If the phantom limb effect allows amputees to produce unique and repeatable sEMG patterns based on the gesture they want to perform, it is conceivable to measure this signal and try to associate it with a precise movement intent. Furthermore, there are other acquisition methods that can either complement or replace sEMG signal to identify a certain gesture.

The forcemyograpy (FMG) uses strain gauges or pressure sensors on the skin's surface to measure the deformation from the underneath muscle [30,31]. This non-invasive method is getting more interest in the transradial prosthesis field as an alternative to the traditional sEMG [32–34]. An arm-band with strain gauges has been designed and could potentially be integrated into an upper-limb prosthesis [35]. In addition to FMG, ultrasound is being regarded as another alternative to sEMG [36] as well as electroencephalography (EEG), which is currently used to command one degree of freedom but presents delay issues [37].

As for invasive methods, while risks involved with the operation are present, they seem to be compensated with remarkable results. Studies delve into the interface with the peripheral nerves [38] or the brain [39] to capture the nerve impulse from the origin. The invasive method that has revolutionized the prosthesis field is the targeted muscle reinnervation (TMR) developed by Kuiken and colleagues [40]. During this operation, the surgeon fixes each loose nerve from the arm to a distinct healthy muscle group, like the dorsal or pectoral muscles. Thus, the response from a certain nerve can be measured at a precise location on the human body. As shown by the arm developed by Johannes and colleagues [41], results from studies using TMR currently surpasses non-invasive methods. However, the risks associated with the operation counteract its own success. A study revealed that TMR can lead to the paralysis of the targeted muscle, the development of a painful neuroma and recurrent pain from the phantom limb [42]. Hence, especially considering the breakthroughs using the phantom limb effect, most of the research groups focus on non-invasive methods to identify the amputee's intent.

5 EMG pattern recognition

We define the intent of movement as the gesture that the amputee wants to achieve. In short, a myoelectric prosthesis tries to identify this intent to activate adequately. The number of degrees of freedom of the prosthesis limits the number of gestures doable by the amputee and we can thus expect the intent to be part of a predefined class.

Classification, or pattern recognition, is a branch of artificial intelligence that associates data to certain classes based on pattern present in these data. This field originates from probability, information theory, and numerical computation that uses machine learning algorithm to train a computer to associate the data to the right class [43].

As seen in Fig. 1.2, a myoelectric prosthesis using pattern recognition would operate as follow: the intent from the amputee first produces electrical activity in the targeted muscle. Then, electrodes matrix around the arm

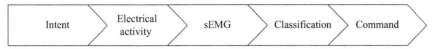

Figure 1.2 *Command steps based on EMG pattern recognition. (Credit: Original)*

captures the sEMG signal from the stump. It serves as input to the classification algorithm that recognizes a certain pattern and associates the signal to a specific gesture. With the intent classified, the right command is sent to the prosthesis.

We define the classification accuracy as the certainty level of associating the intent to the right gesture's class. This accuracy is calculated by inputting a test set to the pattern recognition algorithm and comparing its response with the known intents from the subject. It shows the ratio of good associations across all the test set.

The accuracy is largely used in the field to compare different methods. Although, one issue with this metric is that the studies tend to be highly specific and hard to compare among others. Given that we rarely know what data has been used and how the accuracy has been computed, this ratio often does not represent the same performance from one study to another. For this reason, it has been suggested to build sEMG databases that could be shared and used as benchmarks to assess pattern recognition methods. These databases are currently used for transradial amputees [44], but none exists for the transhumeral population.

Another issue when comparing studies is the type of population used for the tests. The majority of research surrounding myoelectric prosthesis is done on able-bodied subjects. In a 2011 review on hand prosthesis [8], classification accuracies reported for the current solutions were near 90%–95% for the most of them, which would be high enough to justify clinical trials for these prostheses. However, most of them use no or very few amputees to validate their methods. Accuracies for studies using mostly amputees drop to 80%–90%, perhaps even lower. The commonly stated reasons for using able-bodied subjects are the difficulty to gather amputees for research purposes and the ideal development conditions that comes along healthy participants.

Globally, for any classification method, an error rate of 5%–10% can be found between tests on able-bodied or amputees subjects with a lower accuracy for the latter. Interestingly, the performance of a certain method compared to another will be the same among the two populations [2]. Thus, it is not pointless to compare classification methods with an able-bodied population since their relative rank will be equivalent with amputees.

However, to optimize and validate classification models, it should be done with amputees to consider their unique activation pattern and the variability between subjects.

For the past 3 decades, the most used classification methods were machine learning algorithms. These would take features as input and associate them to a class as an output. Given a database for which we know the association between the gestures and the features of the sEMG, a computer can be trained to execute this association. By using different feature sets, altering the parameters, and trying different machine learning methods, it is possible to optimize the algorithm to achieve a better accuracy. In 1993, Hudgins and colleagues [45] demonstrated the possibility to identify a movement intent based on sEMG using machine learning. The time-domain feature set used in this article is still regarded as a golden standard for sEMG pattern recognition and opened the door to a multitude of other methods. Although the accuracy obtained is not sufficient to be clinically viable, this work persists as the standard of the field and studies still tries to surpass it without any significant improvement yet. Apart from time-domain features, frequency-domain features, [46] and wavelet transform has been proposed [47]. The trend of feature engineering has been more popular in the recent years, where teams even create their own features based on observation on the signal and its performance for EMG pattern recognition [48,49].

From usual machine learning methods used for sEMG classification, linear discriminant analysis (LDA) [50] is the most frequently used for its simplicity both in implementation and training [2]. Another popular method is the use of neural network (NN) [1] that seems more suitable for real-time intent recognition [49,51]. Otherwise, support vector machine (SVM) [52] and hidden Markov model (HMM) [53] are also popular among investigated solutions for given feature sets.

Throughout the last years, studies have exploited several known features and machine learning algorithms in an attempt to find a combination that would be significantly better than other proposed methods but failed to do so [2]. There is a general agreement in the field that algorithms such as LDA, NN, SVM, and HMM perform in a very similar way when comparing sEMG classification accuracies. Nevertheless, some research groups still continue to explore usual machine learning algorithms [48] and investigate the possibility of using different features based on the sEMG channel [3].

One major discovery was made while investigating for data that would complete the sEMG information. Known as the "limb position effect" [54], this effect is responsible for altering the sEMG signal based on the arm position, hence sEMG patterns are only repeatable for a given position. Therefore, sEMG data should be captured in several arm positions to include this variation in the training classification dataset and the arm kinematics should be used as a complementary data to determine the intent of the amputee [55]. A study has shown that the error rate decreased by 4.8% when the arm configuration was used as a feature in their NN [56]. To identify this position, research groups often use expensive sensors that require special infrastructure to work, for example, a movement laboratory equipped with infrared cameras. A need has been identified in the literature to move from a laboratory context to a more day-to-day context and to be more mobile [57]. For the measure of the arm position, inertial measurement units (IMU) have been proposed because they are more affordable, compact and they are already frequently used in tandem with sEMG [58].

Considering the stagnant performance from the standard machine learning methods, it is unlikely that these solutions are sufficient to achieve a better control of myoelectric prostheses. The task is even more challenging for transhumeral amputees who have fewer sEMG sites but needs to control even more gestures [59]. It is conceivable that prosthesis control based on sEMG pattern recognition may be the best way to go, but it seems that basic machine learning is not able to take advantage of this signal complexity.

6 Deep learning

Deep learning (DL), another branch of machine learning, is growing fast because of recent progress in computing power and access to big data. This method resolves complex tasks by using several layers of algorithms each with their respective purpose. DL can be seen as a complex mathematical function composed of multiple simple functions (layers) linked between them. Each time data goes through a layer; a new representation of the problem emerges and is input for the next layer [43]. Also, DL methods tend to adapt more easily to variations in the data and eliminates the necessity to decide on a feature set to input like basic machine learning methods would. Indeed, DL can either work with features or the raw sEMG as an input and will learn its own features needed to classify signal.

In the last few years, DL has revolutionized computer vision by achieving lower classification error rates for standard sets when compared to classical machine learning methods [60], and continues to rapidly increase its accuracy for speech recognition [61]. Particularly in vision, DL was able to beat the five last lowest error records with a difference of more than 10% and now takes the lead with an error under 5% [62].

By 2016, Atzori and colleagues [10] presented the use of DL for sEMG-prosthesis applications. The preliminary study concerns the classification of sEMG for hand prosthesis with a convolutional neural network (CNN). Their results showed the possibility to achieve an accuracy close to classical machine learning methods in a short amount of time. Furthermore, their algorithm is able to perform the classification in under 1 ms which bodes well for a myoelectric prosthesis that can work in real-time (<40 ms). They highlight that these results were achieved without any optimization of the parameters and they also identify aspects of their CNN that could be fixed in order to rapidly increase their accuracy.

A review published in 2018 [63] denoted a significant increase in study pairing DL and physiological signals. CNN is still the most studied and popular method for sEMG and other physiological signals, but recent studies showed that RNN might achieve a higher accuracy [64]. They emphasize that DL should be resourceful and easy to implement in a clinical context because this kind of testing often includes a big quantity of data with different types of variation.

Another review from 2018 [65] highlighted the necessity to gather public databases for sEMG because DL algorithms perform best when trained on a large amount of data. Private data owned only by one research group renders the comparison of different studies difficult because often, little is known about the signals used. It also limits the potential of DL methods to achieve greater accuracy by providing only a small number of examples. Since 2016, this problem has been tackled by several (>30) research teams who now offer public data set for physiological signals, including sEMG. Although, for the upper limb, these sEMG databases are almost only gathered on the forearm, with little information about the arm's electrical activity, which is problematic for transhumeral prosthesis development. The same review identified 18 studies tackling sEMG research with DL methods, among which 12 are dedicated to motion recognition. Most of them use healthy subjects while the rest use transradial amputees. Nevertheless, these studies show promising results, better than classical machine learning, for all sEMG pattern recognition application. Besides reaching the desired accuracy in sEMG pattern recognition, the authors highlight the opportunity to assess with DL other issues for prostheses such as electrode shift during usage and muscle fatigue for which machine learning is still subject to.

Among results coming from using DL for myoelectric control, Yamanoi and Kato [66] used a CNN to eliminate the necessity to train the classification algorithm each time the prosthesis is fixed on the user. Standard machine learning methods are sensitive to electrodes location on the arm and can drastically increase their error rate if those are not positioned the same way. This situation is not conceivable for an amputee who needs to put on her/his prosthesis each day. Their CNN showed a better robustness between sessions by maintaining a mean classification accuracy at 57% against 48% for other machine learning methods, hence suggesting CNN might be fitter to absorb those variation coming from placing the electrodes.

Côté-Allard and colleagues [67] also worked with a CNN for hand motion recognition. With a commercial armband, they acquired the sEMG from 35 healthy subjects in order to train their algorithm. A mean accuracy of 97.81% was achieved across all participants and confirms the possibility to use DL on amputees, who should reach an accuracy 5%–10% lower than able-bodied [2], which is encouraging. Furthermore, they present how the algorithm is able to adapt to each subject and how a sEMG dataset could be created for the training of future decision-based methods [68].

7 Usability instead of accuracy

Concerning the control aspect, no specific strategy has been proposed to command a prosthesis based on the output of a classification algorithm. Research groups are currently putting all their efforts to achieve a satisfying motion recognition. For most of the tests involving a virtual or real prosthesis, the decision algorithm either moves the intended device section with a velocity constant or proportional to the intensity of the signal.

It has been identified by researchers and amputees that an unwanted movement is more harmful than having the prosthesis not move at all when a wrong recognition occurs [69]. In other words, when an intent is wrongly identified, it is preferable to not activate the prosthesis rather than executing the wrong command since this command automatically leads to more movements to rectify the trajectory. This situation is frustrating for someone using a prosthesis daily and is part of the reasons for the lack of acceptance of amputees.

Also, a prosthesis is a mechatronic system from a design perspective, but it is considered to be in open-loop and is not autonomous. Its user influences the control of the device by trying to adapt to what she/he is experiencing. In addition to being able to move the rest of her/his body to adjust the trajectory, the amputee is the one deciding on how to reach her/his objective and how to react when an error occurs.

The classification accuracy might be an excellent offline metric to evaluate the success of the sEMG pattern recognition, but it does not guarantee a good control of a prosthesis. To assess the experience of using a certain prosthesis, the term usability has been proposed. It is defined as a measure of how the device is easy to handle to achieve a certain task. Furthermore, because of the harm coming from wrong movements and the amputee's own adjustment, only a weak correlation exists between the accuracy and the usability of a prosthesis [70]. Researchers agree that an accuracy of at least 85% should be achieved before even considering clinical trials.

The same article presents a pair of ratios that better represent the usability of a prosthesis. The first is the total error rate (TER) and computes the number of incorrect decisions divided by the total number of decisions. The second is the active error rate (AER) and computes the number of active incorrect decisions divided by the total number of decisions. Therefore, if an error occurs during the recognition of a motion, moving the prosthesis will increase the AER as not moving the device will not change the ratio, which is what we want to achieve. It should be noted that it is possible to force the AER to 0 by doing no movement each time there is an incorrect decision, which would result in a prosthesis rarely moving. For this reason, AER should always be used with TER [2] since the latter would be high in this situation. Both ratios need to be decreased to achieve a satisfying control of the prosthesis and yet, no standards have been established.

Apart from these metrics, two online tests are used to test the real-time control of a prosthesis control system. The first one was elaborate by Kuiken and colleagues and is called the targeted achievement control (TAC) test [71]. The TAC evaluates the capacity of a participant to start in a certain position and reach an objective in a 3D virtual space. The second test is the Clothespin test [72,73] that asks the participant to grab a clothespin from a string in a 3D virtual space and to put it on another string. Both these tests involve a constant velocity and the time to achieve the task is recorded as the metric of the test and can be compared between different control schemes.

8 Latest trend for transhumeral prosthesis

An examination of the last 5 years was conducted exploring the use of sEMG classification for the transhumeral prosthesis [48,56,74–105]. As seen in Fig. 1.3A, this review shows a steady growth in interest for machine

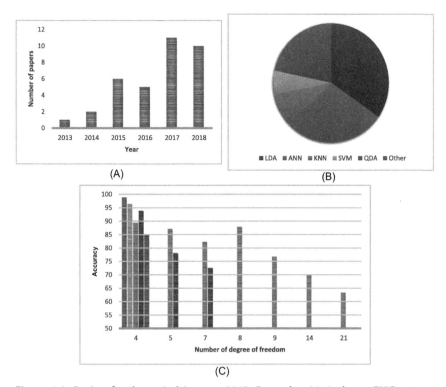

Figure 1.3 *Review for the period January 2013–December 2018 about sEMG pattern recognition for transhumeral prosthesis.* (A) Number per year of studies using machine learning. (B) Machine learning methods employed to classify sEMG. (C) Classification accuracy according to the number of degrees of freedom used in research study. *(Credit: Original)*

learning mirroring the trend in the transradial field. However, this review also indicates that deep learning has not been used for high level of amputation as seen in Fig. 1.3B, where LDA and ANN (classic methods) are the most popular algorithm. It highlights a major difference between the two fields, where transhumeral studies tend to have a delay on the solutions investigated compared to transradial studies.

For the development of transhumeral prosthesis, we look for six degrees of freedom or more to achieve a minimum of functionality. However, as seen in Fig. 1.3C, the accuracies extracted from the review decrease as the number of gestures selected increase, leading to inadequate strategies for amputees. Jarrassé and colleagues [73] presents the only study reaching above 85% for six or more degrees of freedom. They showed great accuracy achieving 88.5% and 86.9% for two transhumeral amputees using a LDA with time domain features. Exceeding the 85% threshold to carry on with a validation stage, they also presented promising results when conducting the clothespin test [72]. This combination of algorithms and features is well known [45] and has been tested often in the past, producing lower accuracies [48,80,101]. Having selected their two participants, this study perhaps shows machine learning is still suitable for certain individuals, but not yet as a general solution.

Regarding the method of investigation, a review of the last 5 years (2013–18 inclusively) was conducted using the following keywords: "upper limb" combined with "arm" and "transhumeral"; "prosthesis" combined with "artificial"; "classification" combined with "recognition"; "myoelectric" combined with "electromyogram," "sEMG," and "EMG". Even though we focus on transhumeral amputees, keywords regarding transradial amputees were not excluded as to ensure articles presenting results for both groups were included. Using two search directories (Compendex, Inspec) and discarding proceedings, letters, and editorials, 34 were discovered and demonstrate the use of EMG pattern recognition with transhumeral amputees.

9 Discussion

Despite high classification accuracies achieved by several studies [75,77,86,91,99], classic methods using machine learning seems inadequate for the control of a prosthesis. Consequently, we report a few papers discussing the implementation of such strategy in a physical device and we instead see several attempts at optimizing the decision algorithm. Among the few teams reaching the validation stage with a virtual or physical prosthesis, insufficient performances were achieved in addition to using able-bodied participants instead of amputees.

Staying at the optimization stage represents an issue because of the low correlation between classification accuracy and the usability of the prosthesis. Although this accuracy is an excellent metric to ascertain the sEMG classification, its optimization does not necessarily lead to a functional device. It is therefore recommended to use ratios like AER/TER that consider how the classification deals with false gestures [2]. It is encouraged to achieve a classification accuracy of 85% with minimal AER/TER ratios for the majority of their participants before moving on to real-time validation.

Also, we highlighted the importance of minimizing the number of false movements from the prosthesis as they are more problematic for the amputee than no movement at all. Surely, a better accuracy will reduce the number of wrong decisions, but other means can be taken to further minimize them. For example, a simple algorithm forcing the rest mode instead of a gesture with low certainty is more advantageous than a system executing every movement.

Furthermore, recent developments concerning transradial prosthesis are now looking into deep learning methods to establish new standards for EMG pattern recognition. A recent review [65] is already showing that a CNN or RNN with a simple architecture exceeds classification accuracies hold by classic methods (LDA, SVM, NN, etc.). This success comes from the capacity of the DL to represent a complex problem with its multiple layers. As for transhumeral studies, while not as advanced or plentiful, they should still initiate this change and invest the necessary resources to acquire a large amount of data and develop DL methods. However, it is worth keeping some progress made in machine learning as they still might apply to new studies. Pre-processing techniques for features and cross-validation are good examples of elements that could enhance a DL method.

To achieve a suitable accuracy with DL, it is necessary to have enough data for its training. The need for public databases that could benefit the entire research community is increasing, especially for transhumeral amputees who still have none at the end of 2018. Databases also provide common data to test a method and directly compare to other studies. Furthermore, by combining data from different populations, gestures, noise sources, pre-processing, and acquisition modality, we give the DL algorithm the means to achieve a more robust classification, a need often mentioned by amputees. Besides, while various techniques allow to work with a limited amount of data, such as data augmentation or transfer learning, they are not as successful.

As of now, no precise type of DL has been able to take the lead regarding EMG pattern recognition. CNN is successful in other areas and can work with various types of data and that's why it is popular in the field. For its part, RNN has a buffer that profit from previous data, making it a powerful tool for a temporal signal like EMG. Furthermore, it is not excluded that an effective solution might be the combination of various types of DL algorithms.

10 Conclusion

To approach and get involved in the design of a transhumeral prosthesis, one must recognize the need for a device that is intuitive to use and stable through its use, that current prostheses on the market fail to be. As for the control based on sEMG pattern recognition, machine learning has been a step forward in the right direction, but a stagnant performance has been observed since the 2000s with no important breakthrough since. The recent developments in artificial intelligence and the access to more complete databases suggest the need for deep learning algorithms. These methods are currently being investigated and show not only the potential to surpass standard machine learning methods, but also to adapt innately to small variations in the signal. Deep learning appears to be the best candidate to achieve a natural and robust control for next-generation prostheses but requires a large amount of data to train. Comparison between offline metrics must be done first, but studies are encouraged to quickly move to clinical trials to evaluate the usability of their solutions. With the help of ratios and online tests, the usability better reflects the capacity of an amputee to control its prosthesis than the commonly used classification accuracy.

References

[1] C.L. Pulliam, J.M. Lambrecht, R.F. Kirsch, EMG-based neural network control of transhumeral prostheses, J. Rehabil. Res. Dev. 48 (6) (2011) 739–754.

[2] E. Scheme, K. Englehart, Electromyogram pattern recognition for control of powered upper-limb prostheses: state of the art and challenges for clinical use, J. Rehabil. Res. Dev. 48 (6) (2011) 643–659.

[3] H.M. Al-Angari, G. Kanitz, S. Tarantino, C. Cipriani, Distance and mutual information methods for EMG feature and channel subset selection for classification of hand movements, Biomed. Signal Process. Control 27 (2016) 24–31.

[4] F. Cordella, et al. Literature review on needs of upper limb prosthesis users, Front. Neurosci. 10 (2016) 209.

[5] W.R. Frontera, J.K. Silver, Fondamenti di medicina fisica e riabilitativa, Verduci, Roma, Italy, (2004).

[6] E. Biddiss, D. Beaton, T. Chau, Consumer design priorities for upper limb prosthetics, Disabil. Rehabil. Assist. Technol. 2 (6) (2007) 346–357.

[7] I.A. Otto, M. Kon, A.H. Schuurman, L.P. van Minnen, Replantation versus prosthetic fitting in traumatic arm amputations: a systematic review, PloS One 10 (9) (2015) e0137729.

[8] B. Peerdeman, et al. Myoelectric forearm prostheses: state of the art from a user-centered perspective, J. Rehabil. Res. Dev. 48 (6) (2011) 719–737.

[9] D. Latour, Prosthetic user-satisfaction and client-centered feedback form, in: Conference Proceedings of the Myoelectric Controls and Upper Limb Prosthetics Symposium, Canada, 2017.

[10] M. Atzori, M. Cognolato, H. Müller, Deep learning with convolutional neural networks applied to electromyography data: a resource for the classification of movements for prosthetic hands, Front. Neurorobotics 10 (2016) 9.

[11] R.J. Pursley, Harness patterns for upper-extremity prostheses, Artif. Limbs 2 (3) (1955) 26–60.

[12] C. Cipriani, F. Zaccone, S. Micera, M.C. Carrozza, On the shared control of an EMG-controlled prosthetic hand: analysis of user #x2013; prosthesis interaction», IEEE Trans. Robot. 24 (1) (2008) 170–184.

[13] Home — Ottobock. [En ligne]. Disponible sur: https://www.ottobock.com/en/. [Consulté le: 25-mai-2018].

[14] Motion Control, Inc., is the leading U.S. manufacturer of myoelectric and externally powered prosthetic arm systems. [En ligne]. Disponible sur: http://www.utaharm.com/. [Consulté le: 25-mai-2018].

[15] LUKE Arm Detail Page, Mobius Bionics. [En ligne]. Disponible sur: http://www.mobiusbionics.com/luke-arm/. [Consulté le: 26-juin-2018].

[16] Coapt, Coapt. [En ligne]. Disponible sur: http://www.coaptengineering.com/. [Consulté le: 25-mai-2018].

[17] R.N. Scott et, P.A. Parker, Myoelectric prostheses: state of the art, J. Med. Eng. Technol. 12 (4) (1988) 143–151.

[18] A.H. Bottomley, Myo-electric control of powered prostheses, J. Bone Joint Surg. Br. 47 (1965) 411–415.

[19] D.S. Dorcas, R.N. Scott, A three-state myo-electric control, Med. Biol. Eng. 4 (4) (1966) 367–370.

[20] A. Fougner, O. Stavdahl, J. Kyberd, Y.G. Losier, P.A. Parker, Control of upper limb prostheses: terminology and proportional myoelectric control-a review, IEEE Trans. Neural Syst. Rehabil. Eng. 20 (5) (2012) 663–677.

[21] H.H. Sears, J. Shaperman, Proportional myoelectric hand control: an evaluation, Am. J. Phys. Med. Rehabil. 70 (1) (1991) 20–28.

[22] L.J. Hargrove, K. Englehart, B. Hudgins, A comparison of surface and intramuscular myoelectric signal classification, IEEE Trans. Biomed. Eng. 54 (5) (2007) 847–853.

[23] G. Drost, D.F. Stegeman, B.G.M. van Engelen, M.J. Zwarts, Clinical applications of high-density surface EMG: a systematic review, J. Electromyogr. Kinesiol. 16 (6) (2006) 586–602.

[24] L.G. Cohen, S. Bandinelli, T.W. Findley, M. Hallett, Motor reorganization after upper limb amputation in man. A study with focal magnetic stimulation, Brain J. Neurol. 114 (Pt. 1B) (1991) 615–627.

[25] V.S. Ramachandran, W. Hirstein, The perception of phantom limbs. The D. O. Hebb lecture, Brain J. Neurol. 121 (Pt 9) (1998) 1603–1630.

[26] C.M. Kooijman, P.U. Dijkstra, J.H. Geertzen, A. Elzinga, C.P. van der Schans, Phantom pain and phantom sensations in upper limb amputees: an epidemiological study, Pain 87 (1) (2000) 33–41.

[27] K.T. Reilly, C. Mercier, M.H. Schieber, A. Sirigu, Persistent hand motor commands in the amputees' brain, Brain J. Neurol. 129 (Pt. 8) (2006) 2211–2223.

[28] G. Gaudet, M. Raison, F.D. Maso, S. Achiche, M. Begon, Intra- and intersession reliability of surface electromyography on muscles actuating the forearm during maximum voluntary contractions, J. Appl. Biomech. 32 (6) (2016) 558–570.

[29] P.A. Parker, K.B. Englehart, B.S. Hudgins, Control of Powered Upper Limb Prostheses, in Electromyography: Physiology, Engineering, and Non-invasive Applications, John Wiley & Sons, Hoboken, New Jersey, 2004, pp. 453–475.

[30] M. Wininger, N.-H. Kim, W. Craelius, Pressure signature of forearm as predictor of grip force, J. Rehabil. Res. Dev. 45 (6) (2008) 883–892.

[31] C. Zizoua, M. Raison, S. Boukhenous, M. Attari, S. Achiche, Detecting muscle contractions using strain gauges, Electron. Lett. 52 (22) (2016) 1836–1838.

[32] E. Cho, R. Chen, L.-K. Merhi, Z. Xiao, B. Pousett, C. Menon, Force myography to control robotic upper extremity prostheses: a feasibility study, Front. Bioeng. Biotechnol. 4 (2016) 18.

[33] X. Jiang, L.-K. Merhi, Z.G. Xiao, C. Menon, Exploration of force myography and surface electromyography in hand gesture classification », Med. Eng. Phys. 41 (2017) 63–73.

[34] A. Radmand, E. Scheme, K. Englehart, High-density force myography: a possible alternative for upper-limb prosthetic control, J. Rehabil. Res. Dev. 53 (4) (2016) 443–456.

[35] C. Zizoua, M. Raison, S. Boukhenous, M. Attari, S. Achiche, Development of a bracelet with strain-gauge matrix for movement intention identification in traumatic amputees, IEEE Sens. J. 17 (8) (2017) 2464–2471.

[36] D. Sierra González, C. Castellini, A realistic implementation of ultrasound imaging as a human-machine interface for upper-limb amputees, Front. Neurorobotics 7 (2013) 17.

[37] T. Beyrouthy, S.K.A. Kork, J.A. Korbane, A. Abdulmonem, EEG Mind controlled Smart Prosthetic Arm, in: 2016 IEEE International Conference on Emerging Technologies and Innovative Business Practices for the Transformation of Societies (EmergiTech), 2016, pp. 404–409.

[38] M.G. Urbanchek, Z. Baghmanli, J.D. Moon, K.B. Sugg, N.B. Langhals, P.S. Cederna, Quantification of regenerative peripheral nerve interface signal transmission, Plast. Reconstr. Surg. 130 (2012) 55–56.

[39] C.A. Chestek, et al. Long-term stability of neural prosthetic control signals from silicon cortical arrays in rhesus macaque motor cortex, J. Neural Eng. 8 (4) (2011) 045005.

[40] K.D. O'Shaughnessy, G.A. Dumanian, R.D. Lipschutz, L.A. Miller, K. Stubblefield, T.A. Kuiken, Targeted reinnervation to improve prosthesis control in transhumeral amputees. A report of three cases, J. Bone Joint Surg. Am. 90 (2) (2008) 393–400.

[41] M.S. Johannes, J.D. Bigelow, J.M. Burck, S.D. Harshbarger, T. Van Doren, An overview of the developmental process for the modular prosthetic limb, John Hopkins APL Tech. Dig. 30 (3) (2011) 207–216.

[42] T.A. Kuiken, et al. Targeted reinnervation for enhanced prosthetic arm function in a woman with a proximal amputation: a case study, Lancet Lond. Engl. 369 (2007) 371–380 9559.

[43] I. Goodfellow, Y. Bengio, A. Courville, Deep Learning, MIT Press, Cambridge, MA, (2016).

[44] M. Atzori, et al. Characterization of a benchmark database for myoelectric movement classification, IEEE Trans. Neural Syst. Rehabil. Eng. 23 (1) (2015) 73–83.

[45] B. Hudgins, S. Parker, R.N. Scott, A new strategy for multifunction myoelectric control, IEEE Trans. Biomed. Eng. 40 (1) (1993) 82–94.

[46] K. Englehart, B. Hudgins, P.A. Parker, M. Stevenson, Classification of the myoelectric signal using time-frequency based representations, Med. Eng. Phys., 21, (6–7), (1999) 431–438.

[47] A. Phinyomark, C. Limsakul, P. Phukpattaranont, Application of wavelet analysis in EMG feature extraction for pattern classification, Meas. Sci. Rev. 11 (2) (2011) 45–52.

[48] O.W. Samuel, et al. Pattern recognition of electromyography signals based on novel time domain features for amputees' limb motion classification, Comput. Electr. Eng. 67 (2018) 646–655.

[49] S. Guo, M. Pang, B. Gao, H. Hirata, H. Ishihara, Comparison of sEMG-based feature extraction and motion classification methods for upper-limb movement, Sensors 15 (4) (2015) 9022–9038.

[50] K. Englehart, B. Hudgins, A robust, real-time control scheme for multifunction myoelectric control, IEEE Trans. Biomed. Eng. 50 (7) (2003) 848–854.

[51] S. Guo, M. Pang, Y. Sugi, Y. Nakatsuka, Study on the comparison of three different upper limb motion recognition methods, in: 2014 IEEE International Conference on Information and Automation (ICIA), 2014, pp. 208–212.

[52] A. Ameri, E.N. Kamavuako, E.J. Scheme, K.B. Englehart, P.A. Parker, Support vector regression for improved real-time, simultaneous myoelectric control, IEEE Trans. Neural Syst. Rehabil. Eng. 22 (6) (2014) 1198–1209.

[53] A.D.C. Chan et, K.B. Englehart, Continuous myoelectric control for powered prostheses using hidden Markov models, IEEE Trans. Biomed. Eng. 52 (1) (2005) 121–124.

[54] A. Fougner, E. Scheme, A.D.C. Chan, K. Englehart, Ø. Stavdahl, Resolving the limb position effect in myoelectric pattern recognition, IEEE Trans. Neural Syst. Rehabil. Eng. 19 (6) (2011) 644–651.

[55] A. Radmand, E. Scheme, K. Englehart, On the suitability of integrating accelerometry data with electromyography signals for resolving the effect of changes in limb position during dynamic limb movement, J. Prosthet. Orthot. 26 (4) (2014) 185–193.

[56] G. Gaudet, M. Raison, S. Achiche, Classification of upper limb phantom movements in transhumeral amputees using electromyographic and kinematic features, Eng. Appl. Artif. Intell. 68 (2018) 153–164.

[57] S. Hernandez, M. Raison, A. Torres, G. Gaudet, S. Achiche, From on-body sensors to in-body data for health monitoring and medical robotics: a survey, in: 2014 Global Information Infrastructure and Networking Symposium (GIIS), 2014, pp. 1–5.

[58] A. Duivenvoorden, K. Lee, M. Raison, S. Achiche, Sensor fusion in upper limb area networks: a survey, in: 2017 Global Information Infrastructure and Networking Symposium (GIIS), 2017, pp. 56–63.

[59] P. Pilarski, A.L. Edwards, K.M. et Chan, Novel control strategies for arm prostheses: a partnership between man and machine, Jpn. J. Rehabil. Med. 52 (2) (2015) 91–95.

[60] A. Krizhevsky, I. Sutskever, G.E. Hinton, ImageNet classification with deep convolutional neural networks, Commun. ACM 60 (6) (2017) 84–90.

[61] L. Deng, G. Hinton, B. Kingsbury, New types of deep neural network learning for speech recognition and related applications: an overview, in: 2013 IEEE International Conference on Acoustics, Speech and Signal Processing, 2013, pp. 8599–8603.

[62] K. He, X. Zhang, S. Ren, J. Sun, Delving deep into rectifiers: surpassing human-level performance on imagenet classification, 2015, arXiv:1502.01852[cs.CV].

[63] O. Faust, Y. Hagiwara, T.J. Hong, O.S. Lih, U.R. Acharya, Deep learning for healthcare applications based on physiological signals: a review, Comput. Methods Programs Biomed. 161 (2018) 1–13.

[64] R. Laezza, Deep neural networks for myoelectric pattern recognition—an implementation for multifunctional control., Master's Thesis, Chalmers University of Technologies, Gothenburg, Sweden, 2018.

[65] A. Phinyomark, E. Scheme, A. Phinyomark, E. Scheme, EMG pattern recognition in the era of big data and deep learning, Big Data Cogn. Comput. 2 (3) (2018) 21.

[66] Y. Yamanoi et R. Kato, Control method for myoelectric hand using convolutional neural network to simplify learning of EMG signals, in: 2017 IEEE International Conference on Cyborg and Bionic Systems (CBS), 2017, pp. 114–118.

[67] U.C. Allard et al., A convolutional neural network for robotic arm guidance using sEMG based frequency-features, in: 2016 IEEE/RSJ International Conference on Intelligent Robots and Systems (IROS), 2016, pp. 2464–2470.

[68] U. Côté-Allard, C.L. Fall, A. Campeau-Lecours, C. Gosselin, F. Laviolette, B. Gosselin, Transfer learning for sEMG hand gestures recognition using convolutional neural networks, in: 2017 IEEE International Conference on Systems, Man, and Cybernetics (SMC), 2017, 1663–1668.

[69] L. Hargrove, Y. Losier, B. Lock, K. Englehart, B. Hudgins, A real-time pattern recognition based myoelectric control usability study implemented in a virtual environment, in: 2007 29th Annual International Conference of the IEEE Engineering in Medicine and Biology Society, 2007, 4842–4845.

[70] B.A. Lock, K. Englehart, B. Hudgins, Real-time myoeletric control in a virtual environment to relate usability vs. accuracy, in: MEC Symposium Conference Proceedings, Fredericton, Canada, 2005.

[71] A.M. Simon, L.J. Hargrove, B.A. Lock, T.A. Kuiken, Target achievement control test: evaluating real-time myoelectric pattern-recognition control of multifunctional upper-limb prostheses, J. Rehabil. Res. Dev. 48 (6) (2011) 619–627.

[72] P. Kyberd, A. Hussaini, G. Maillet, Characterisation of the clothespin relocation test as a functional assessment tool, J. Rehabil. Assist. Technol. Eng. 5 (2018) 2055668317750810.

[73] A. Hussaini, P. Kyberd, Refined clothespin relocation test and assessment of motion, Prosthet. Orthot. Int. 41 (3) (2017) 294–302.

[74] K. Bakshi, R. Pramanik, M. Manjunatha, et C. S. Kumar, Upper limb prosthesis control: a hybrid eeg-emg scheme for motion estimation in transhumeral subjects, in: 2018 40th Annual International Conference of the IEEE Engineering in Medicine and Biology Society (EMBC), Honolulu, HI, 2018, pp. 2024–2027.

[75] N. Jarrassé, et al. Phantom-mobility-based prosthesis control in transhumeral amputees without surgical reinnervation: a preliminary study, Front. Bioeng. Biotechnol. 6 (2018) 164.

[76] N. Jarrassé, et al. Classification of phantom finger, hand, wrist, and elbow voluntary gestures in transhumeral amputees with sEMG, IEEE Trans. Neural Syst. Rehabil. Eng. 25 (1) (2017) 71–80.

[77] H.I. Aly, S. Youssef, C. Fathy, Hybrid brain computer interface for movement control of upper limb prostheses, in: 2018 International Conference on Biomedical Engineering and Applications (ICBEA), Funchal, 2018, pp. 1–6.

[78] Y. Ogiri, Y. Yamanoi, W. Nishino, R. Kato, T. Takagi, H. Yokoi, Development of an upper limb neuroprosthesis to voluntarily control elbow and hand, in: 2017 26th IEEE International Symposium on Robot and Human Interactive Communication (RO-MAN), 2017, pp. 298–303.

[79] D.S.V. Bandara, J. Arata, K. Kiguchi, Towards control of a transhumeral prosthesis with EEG signals, Bioengineering 5 (2) (2018) 26.

[80] D. Bai et al., Intelligent prosthetic arm force control based on sEMG analysis and BPNN classifier, 2017 IEEE Int. Conf. Cyborg Bionic Syst. CBS, 2017, 108–113.

[81] J. Wang, W. Wichakool, Artificial elbow joint classification using upper arm based on surface-EMG signal, in: 2017 IEEE 3rd International Conference on Engineering Technologies and Social Sciences (ICETSS), 2017, pp. 1–4.

[82] S. Herle, Movement intention detection from SEMG signals using time-domain features and discriminant analysis classifiers, in: 2018 IEEE International Conference on Automation, Quality and Testing, Robotics (AQTR), Cluj-Napoca, 2018, pp. 1–6.

[83] K. Veer and T. Sharma, Electromyographic classification of effort in muscle strength assessment, Biomed. Eng. Biomed. Tech., 63 (2) (2018) 131–137.

[84] Y. Xu, D. Zhang, Y. Wang, J. Feng, W. Xu, Two ways to improve myoelectric control for a transhumeral amputee after targeted muscle reinnervation: a case study, J. NeuroEngineering Rehabil. 15 (1) (2018) 37.

[85] K.R. Lyons, S.S. Joshi, Upper limb prosthesis control for high-level amputees via myoelectric recognition of leg gestures, IEEE Trans. Neural Syst. Rehabil. Eng. 26 (5) (2018) 1056–1066.

[86] K. Veer, R. Vig, Analysis and recognition of operations using SEMG from upper arm muscles, Expert Syst. 34 (6) (2017) e12221.

[87] N. Jose, R. Raj, P.K. Adithya, K.S. Sivanadan, Classification of forearm movements from sEMG time domain features using machine learning algorithms, in: TENCON 2017-2017 IEEE Region 10 Conference, 2017, pp. 1624–1628.

[88] N. Jarrassé et al., Voluntary phantom hand and finger movements in transhumerai amputees could be used to naturally control polydigital prostheses, in: 2017 International Conference on Rehabilitation Robotics (ICORR), 2017, pp. 1239–1245.

[89] L. He, P.A. Mathieu, Muscle synergy of biceps brachii and online classification of upper limb posture, in: 2017 International Conference on Virtual Rehabilitation (ICVR), 2017, pp. 1–6.

[90] Q. Wu, J. Shao, X. Wu, Y. Zhou, F. Liu, F. Xiao, Upper limb motion recognition based on LLE-ELM method of sEMG, Int. J. Pattern Recognit. Artif. Intell. 31 (06) (2016) 1750018.

[91] X. Li, O.W. Samuel, X. Zhang, H. Wang, P. Fang, G. Li, A motion-classification strategy based on sEMG-EEG signal combination for upper-limb amputees, J. NeuroEngineering Rehabil., 14 (2017).

[92] M. Salama, A. Bakr, Six prosthetic arm movements using electromyogram signals: a prototype, in: 2016 UKSim-AMSS 18th International Conference on Computer Modelling and Simulation (UKSim), 2016, pp. 37–42.

[93] T. Lenzi, J. Lipsey, J.W. Sensinger, The RIC arm—a small anthropomorphic transhumeral prosthesis, IEEEASME Trans. Mechatron., 21 6 (2016) 2660–2671.

[94] A. Ehrampoosh, A. Yousefi-koma, M. Ayati, Development of myoelectric interface based on pattern recognition and regression based models, in: 2016 Artificial Intelligence and Robotics (IRANOPEN), 2016,pp. 145–150.

[95] C.W. Antuvan, F. Bisio, F. Marini, S.-C. Yen, E. Cambria, L. Masia, Role of muscle synergies in real-time classification of upper limb motions using extreme learning machines, J. NeuroEngineering Rehabil. 13 (2016) 76.

[96] K. Veer, T. Sharma, R. Agarwal, A neural network-based electromyography motion classifier for upper limb activities, J. Innov. Opt. Health Sci. 09 (06) (2016) 1650025.

[97] P. Azaripasand, A. Maleki, A. Fallah, Classification of ADLs using muscle activation waveform versus thirteen EMG features, in: 2015 22nd Iranian Conference on Biomedical Engineering (ICBME), 2015, pp. 189–193.

[98] Q. Zhang, C. Zheng, C. Xiong, EMG-based estimation of shoulder and elbow joint angles for intuitive myoelectric control, in: 2015 IEEE International Conference on Cyber Technology in Automation, Control, and Intelligent Systems (CYBER), 2015, pp. 1912–1916.

[99] J. Gade, R. Hugosdottir, E. N. Kamavuako, Phantom movements from physiologically inappropriate muscles: A case study with a high transhumeral amputee, in: 2015 37th Annual International Conference of the IEEE Engineering in Medicine and Biology Society (EMBC), 2015, 3488–3491.

[100] C.W. Antuvan, F. Bisio, E. Cambria, L. Masia, Muscle synergies for reliable classification of arm motions using myoelectric interface, in: 2015 37th Annual International Conference of the IEEE Engineering in Medicine and Biology Society (EMBC), 2015, pp. 1136–1139.

[101] K. Veer, A technique for classification and decomposition of muscle signal for control of myoelectric prostheses based on wavelet statistical classifier, Measurement 60 (2015) 283–291.

[102] A. Suberbiola, E. Zulueta, J.M. Lopez-Guede, I. Etxeberria-Agiriano, M. Graña, Arm orthosis/prosthesis movement control based on surface EMG signal extraction, Int. J. Neural Syst. 25 (03) (2015) 1550009.

[103] R. Almada-Aguilar, L.M. Torres-Treviño, G. Quiroz, Establishing a simplified functional relationship between EMG signals and actuation signals using artificial neural networks, in: 2014 13th Mexican International Conference on Artificial Intelligence, 2014, pp. 128–132.

[104] D.C. Tkach, A.J. Young, L.H. Smith, E.J. Rouse, L.J. Hargrove, Real-time and offline performance of pattern recognition myoelectric control using a generic electrode grid with targeted muscle reinnervation patients, IEEE Trans. Neural Syst. Rehabil. Eng. 22 (4) (2014) 727–734.

[105] L.J. Hargrove, B.A. Lock, A.M. Simon, Pattern recognition control outperforms conventional myoelectric control in upper limb patients with targeted muscle reinnervation, in: 2013 35th Annual International Conference of the IEEE Engineering in Medicine and Biology Society (EMBC), 2013, pp. 1599–1602.

CHAPTER 2

The 2-DOF mechanical impedance of the human ankle during poses of the stance phase

Guilherme A. Ribeiro, Lauren N. Knop, Mo Rastgaar[1]
Polytechnic Institute, Purdue University, West Lafayette, IN, United States

1 Introduction

The ankle is the first major joint that interacts with the environment, transferring the ground reaction forces, and torques to the remaining of the body. In particular, for transtibial amputees, the accurate characterization of the healthy human ankle behavior is essential to the design of safe and robust powered prosthesis controllers, in the way it interacts with the environment and with the user. When an environmental torque or force acts on the prosthetic foot, the ankle must react with a torque against this disturbance and decrease the discomfort to the user, ideally the same way a healthy ankle would do. This reaction torque to an ankle motion disturbance is defined as the mechanical impedance of the human ankle and is the subject of study for many decades.

Weiss and Kearney estimated the ankle impedance along the Dorsi-Plantar (DP) motion as a 2nd order dynamic model and related it to varying levels of ankle angle and torque for subjects lying in the supine position [1,2]. Also, with an unloaded ankle condition, Lee showed that the impedance has an anisotropic behavior, having a different response for a different axis of rotation and muscle contraction levels [3,4]. Tehrani looked at calf muscles as Hill-type muscles to proposed that the ankle impedance, commonly modeled as a 2nd order system, is better described as a 3rd order, with three poles and two zeros [5]. Recently, the stiffness component of the impedance was found to increase when the center of pressure (CoP) of the ground reaction force moves along the anterior direction [6,7]. As the standing subject naturally swayed or purposely leaned forward, ground

[1] Contact Author: rastgaar@purdue.edu.
Research supported by NSF grants 1921046 and 1923760.

perturbations excited the ankle to identify the stiffness in correlation to the CoP.

This finding agrees with another branch of impedance estimation that studies the functional DP mechanical impedance during the gait cycle, in which it was found to change over time. This time-varying characteristic was found across all the sub-phases of the gait: the stance, in which the stiffness was shown to increase over time [8,9]; the swing [10], with lower stiffness and damping values; and the terminal stance [11], showing a transitioning stiffness and a spike on the damping. The modulation of impedance across time was also verified in others DoF, during straight walk [8], and turning maneuvers [12]. There may be many more different ankle impedance curves for other conditions, such as varying walking speeds, terrain inclines, stairs ascent/descent. Therefore, this study aims to generalize the behavior of the mechanical impedance by correlating it to the generalized variable, CoP, comparing to the results of functional experiments.

In this work, a 2nd order ankle impedance was estimated in the DP and inversion-eversion (IE) directions while the subjects stood in four static poses, resembling stages of the gait cycle. The dynamics of the apparatus and the foot inertia was also estimated and compensated to reduce its effects on the dynamic response of the ankle. In addition, a numerical simulation was developed to verify the correctness and accuracy of the proposed method, showing promising results. Interestingly, the estimated ankle impedance showed results similar to functional experiments. This fact may have important implications for the protocol design of new ankle impedance estimation experiments.

2 Methods

2.1 Subjects

This study included 15 male subjects with no previously reported musculoskeletal injuries. Each subject gave written consent to participate in this study through the Michigan Technological University Institutional Review Board (IRB), where the experiments were conducted. Table 2.1 includes the mean and standard deviation of the physiological information across the subject population; including the age, mass, height, and foot length (FL).

Table 2.1 Subject physiological data.

Age	Mass (kg)	Height (m)	Foot length (cm)
28 ± 4.3	79 ± 10.7	1.8 ± 0.07	26.5 ± 1.4

2.2 Experimental setup

The ankle impedance was estimated using an established instrumented walkway that can measure the ankle angle displacement and ground reaction torques while the ankle is subjected to perturbations, as described in Ref. [13]. As shown in Fig. 2.1, the instrumented walkway consists of a motion capture system (8—Optitrack 17W cameras, 350 fps) and a force plate (Kistler 9260AA3, 3500 Hz sampling rate) mounted to a vibrating platform. When actuated, the vibrating platform rotates around a universal joint, located below the ground plane, generating motion in the vertical planes (a combination of motion in the frontal and sagittal planes).

In addition, EMG sensors (Delsys Trigno wireless, 2000 Hz sampling rate) were used to measure the lower extremity muscle activity. The sensors were placed on five muscles; including the tibialis anterior (TA), the peroneus longus (PL), the soleus (SOL), the gastrocnemius lateral (GAL), and the gastrocnemius medial (GAM). These muscles were selected based on their antagonistic properties and their contribution to motion in the DP and IE axes (cite).

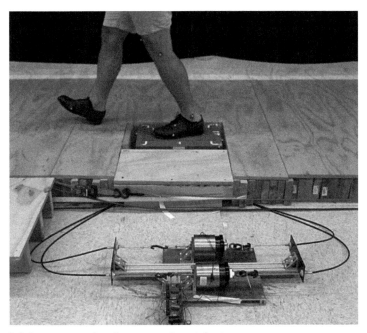

Figure 2.1 Instrumented walkway experimental set up.

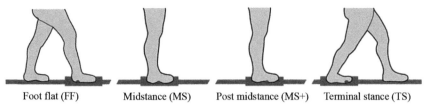

| Foot flat (FF) | Midstance (MS) | Post midstance (MS+) | Terminal stance (TS) |

Figure 2.2 Static gait poses emulating 4 sub-phases of the gait cycle: foot flat (FF), midstance (MS), post midstance (MS+), and terminal stance (TS). The *red dot* represents the COP position.

2.3 Experimental protocol

Four standing poses were analyzed in this study: foot flat (FF), midstance (MS), post midstance (MS+), and terminal stance (TS), as shown in Fig. 2.2. These standing poses varied the subject's ankle angle and the center of pressure of the foot, and were selected to emulate the ankle angle and COP during the stance phase of walking (cite). However, different from walking, these poses remained static through each test. Three rounds of trials were repeated, within each, the four poses were experimented in random order, totaling 12 trials. To avoid fatigue, each trial lasted 30 s, and at least 1-min interval was enforced between trials.

To have consistency between repeated trials, the placement of the feet, the COP position, and the weight distribution were supervised. The right foot was placed on the force plate inside a drawn outline of the foot, while the left foot was behind (FF), aligned (MS, MS+), or in front of the right foot. For the FF and TS poses, the stance length (anterior–posterior distance between the feet) was defined as 40% of the subjects' height. Lastly, the subjects relied on real-time measurements on a monitor screen to maintain the COP of the perturbed foot on target and hold equal weight support between both legs. The target COP locations were 30.6%, 40.5%, 53.0%, and 63.6% of the foot length, from the heel, for poses FF, MS, MS+, and TS, respectively.

The ground perturbations were in the form of pulse trains of random, rotating axis (0–360°), period (0.03–0.2 s), and duration (0.9–1.1 s); and between consecutive pulse train, the vibrating platform was inactive for a random pause time (0.9–1.1 s), as shown in Fig. 2.3. The use of random durations and pause periods was intended to decrease reflex responses and predictive muscle contraction by the subjects.

2.4 Dynamic simulation with Simulink multibody

To validate the identification algorithm, in special the correctness of the differential equations described in later sections, a dynamic simulation on Simulink multibody was developed (Fig. 2.4). This simulation replicated the

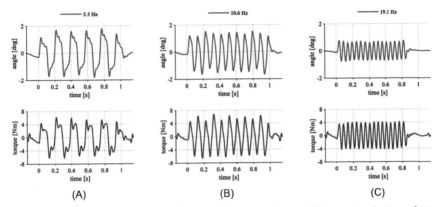

Figure 2.3 Examples of ankle angle (top) and torque (bottom) for a perturbation of (A) 5.5, (B) 10.6, and (C) 19.1 Hz. Perturbation started at 0 s and lasted for 0.9–1.1 s.

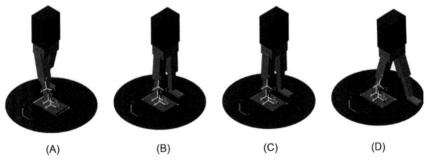

Figure 2.4 Dynamic simulation of the experimental setup for different foot placements. The simulation was developed to test possible artifacts on the ankle impedance due to upper body dynamics, apparatus inertial, mean torque and angle, affecting the bias and variance of the estimates.

same experimental protocol (perturbation type, pose repetitions, COP, and tracking of weight distribution) to account for possible algorithmic artifacts. In addition, it replicated the kinematic constraints and output power of the vibrating platform.

The simulated subjects had a foot limb moment of inertia, the product of inertia, mass, and center of mass vector of $J_F = [0.7, \ 3.4, \ 3.0]^T$ gm^2, $P_F = [0,0,0]^T$ gm^2, $m_F = 0.84$ kg, and $r_F = [15, \ -20, \ 0]^T$ mm, respectively, comparable to experimental foot inertia estimates [14]. The FP inertia parameters were set to the average of the experimental estimates (presented in later sections). The measurement noise followed a normal distribution with zero mean and standard deviation of 0.1 N, 0.05 Nm, 0.006°, 0.03°, 0.04°, 0.02 mm, 0.02 mm, and 0.02 mm for the force, torque, orientation of FP, foot, and shank, and translation of FP, foot, and shank, respectively.

These noise parameters were determined from the experimental tests by matching the power spectrum density (PSD) of the measurements at higher frequency ranges, where the measurement noise is more dominant.

Thirty artificial subjects were simulated with constant ankle stiffness and damping selected from a uniform random distribution ranging from $[0,0,0]^T$ to $[160, 550, 550]^T$ Nm/rad for stiffness and $[-0.8, -0.8, -0.8]^T$ to $[1.6, 1.6, 1.6]^T$ Nms/rad for damping, to cover the full range of impedance found from experiments with human subjects (described in later sections). Each of these simulated subjects was tested for each pose, totaling 120 simulations of approximately 50 perturbations each. Differently from the experimental tests, in the simulation tests, the accuracy of the parameter estimation can be assessed because the nominal impedance is known.

2.5 Analysis methods

2.5.1 Apparatus dynamics

The ground force and torque react to the subjects' ankle dynamics but also reacts to the apparatus dynamics (force plate inertia and bias noise). To estimate the unbiased ankle impedance, the apparatus dynamics can be identified in an additional experiment in which the vibrating platform is actuated freely, without contact with the subjects. In this experiment, the unloaded vibrating platform was actuated with a stochastic signal (30 Hz update rate) for 30 s. The ground reaction torque and force can be modeled as in Eq. (2.1) and (2.2), respectively:

$$_0T_P = I_p\dot{\omega}_p + \omega_p \times \left(I_p\omega_p\right) + r_p \times \left[m_p\left(\ddot{p}_{P0} - g\right)\right] - \dot{p}_p \times \left(m_p\dot{p}_{P0}\right) + T_{bias} \quad (2.1)$$

$$_0F_P = m_p\left(\ddot{p}_{P0} - g\right) + F_{bias} \quad (2.2)$$

where $m_p, r_p,$ and I_p are the inertial parameters of the FP: the mass, center of mass relative to the FP origin, and the inertia matrix about the center of mass; $\omega_p, \dot{\omega}_p, \dot{p}_p, p_P, \dot{p}_{P0} = \dot{p}_p + \omega_p \times r_p,$ and $\ddot{p}_{P0} = \ddot{p}_p + \dot{\omega}_p \times r_p + \omega_p \times \left(\omega_p \times r_p\right)$ are the angular velocity and acceleration about the body, linear velocity and acceleration of FP origin, linear velocity and acceleration of the FP center of mass (Fig. 2.5); $T_{bias}, F_{bias},$ are the constant torque and force due to sensor zeroing and bias noise, and $g = [0, -9.81, 0]^T$ m/s² is the gravity vector, respectively.

The unknown parameters from Eqs. (2.1) and (2.2) are estimated from a calibration experiment and later used to compensate the ground reaction forces and torques from the human experiments. This estimation is a problem of non-linear optimization of the form

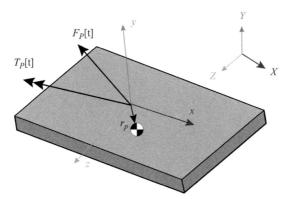

Figure 2.5 Coordinate frame notation for the FP. Origin of the body lies in the center of the top surface, with x point toward the long body dimension and y normal to the top surface. The measured force and torque, $T_p[t]$ and $F_p[t]$, act on the origin of the body.

$$\theta^\star \triangleq \underset{\theta}{\arg\min} \sum_t \left(\varepsilon_\theta[t]\right)^T \varepsilon_\theta[t] \tag{2.3}$$

where

$$\varepsilon_\theta[t] \triangleq \begin{bmatrix} I_p\dot{\omega}_p[t] + \omega_p[t] \times \left(I_p\omega_p[t]\right) + r_p \times \left(m_p\left(\ddot{p}_{P0}[t] - g\right)\right) \\ -\dot{p}_p[t] \times \left(m_p\dot{p}_{P0}[t]\right) + T_{bias} - T_p[t] \\ m_p\left(\ddot{p}_{P0}[t] - g\right) + F_{bias} - F_p[t] \end{bmatrix} \in \mathbb{R}^6 \tag{2.4}$$

$$\theta \triangleq \begin{bmatrix} flat\left(I_p\right)^T & m_p & r_p^T & T_{bias}^T & F_{bias}^T \end{bmatrix}^T \in \mathbb{R}^{16} \tag{2.5}$$

are the cost function [derived from the residual of Eqs. (2.1) and (2.2)] and vector of unknown parameters, respectively. The operators $\star[t]$ and $flat(\star): \mathbb{R}^{3\times3} \to \mathbb{R}^6$ represent a measurement from time t and the conversion between an inertia matrix to a column vector with the 3 moments of inertia and 3 products of inertia, respectively.

Signal processing. The time derivatives were numerically computed via Savitzky-Golay filter [15] (5th order polynomial in a 15–samples window) and the optimization solved with MATLAB's *fmincon* function (interior-point algorithm [16]).

2.5.2 Ankle impedance estimation
Similar to Eq. 1 and 2, the torque due to the foot inertia can be modeled as

$$_0T_F = I_F\dot{\omega}_F + \omega_F \times \left(I_F\omega_F\right) + r_F \times \left[m_F\left(\ddot{p}_{F0} - g\right)\right] - \dot{p}_F \times \left(m_F\dot{p}_{F0}\right) \tag{2.6}$$

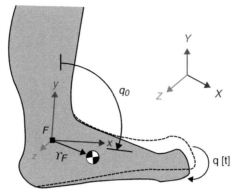

Figure 2.6 Coordinate frame notation for the foot. Foot origin is on ankle joint, with x pointing toward the long axis of the foot, parallel to the ground, and y pointing upward. The orientation of the foot in respect to the shank is composed by a mean angle q_0 and a small rotation $q[t]$ due to the ground perturbation.

$$\dot{p}_{F0} = \dot{p}_F + \omega_F \times r_F \tag{2.7}$$

$$\ddot{p}_{F0} = \ddot{p}_F + \dot{\omega}_F \times r_F + \omega_F \times \left(\omega_F \times r_F\right) \tag{2.8}$$

where m_F, r_F, and I_F are the inertial parameters of the foot: the mass, center of mass relative to the foot origin (ankle), and the inertia matrix about the center of mass; ω_F, $\dot{\omega}_F$, $\dot{p}_F$, $\ddot{p}_F$, $\dot{p}_{F0} = \dot{p}_F + \omega_F \times r_F$, and $\ddot{p}_{F0} = \ddot{p}_F + \dot{\omega}_F \times r_F + \omega_F \times \left(\omega_F \times r_F\right)$ are the angular velocity and acceleration about the body, linear velocity and acceleration of foot origin, linear velocity, and acceleration of the foot center of mass (Fig. 2.6).

The external torque acting on the ankle, compensating for the torque component acting on the FP inertia, is calculated as

$$T_F[t] \triangleq T_P[t] - {}_0 T_P[t] + \left(p_P[t] - p_F[t]\right) \times \left(F_P[t] - {}_0 F_P[t]\right) \tag{2.9}$$

The orientation between the shin and foot, $R_F[t] = R_S[t]R_{q_0}R_q[t]$, can be decomposed in two transformations: the initial (neutral) orientation, $R_{q_0} \in SO(3)$, at a time t_0, before the perturbation; and a smaller angular displacement, $R_q[t] \in SO(3)$, due to the perturbation. The impedance is estimated based on this small angular displacement, $q[t]$ (Fig. 2.7A).

$$R_{q_0} = R_S^T\left[t_0\right]R_F\left[t_0\right] \tag{2.10}$$

$$R_q[t] = R_{q_0}^T R_S^T[t]R_F[t] \tag{2.11}$$

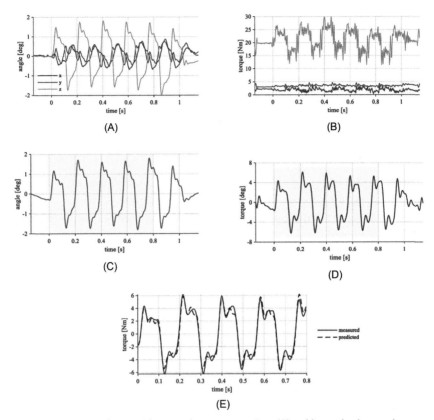

Figure 2.7 Stages of the ankle impedance estimation. (A) ankle angle due to the perturbation, (B) input torque acting on ankle without the FP inertia torques, (C–D) filtered angles and torques, with the region of interest highlighted, and (E) the prediction of the torque using kinematics of the ankle system and estimated impedance parameters.

The DP and IE ankle angles are defined as the z and x rotations of the Euler angle (XYZ) representation of $R_q[t]$. The conversion between the rotation matrix to Euler angles is presented as

$$q[t] = \left[\ -\mathrm{atan}\left(R_q^{2,3}[t]/R_q^{3,3}[t]\right) \quad \mathrm{asin}\left(R_q^{2,3}[t]\right) \quad -\mathrm{atan}\left(R_q^{1,2}[t]/R_q^{1,1}[t]\right) \ \right]^T$$

$$(2.12)$$

Finally, the ankle impedance was defined as a 2nd order model for DP and IE ankle motions, relating the small angle displacements to ankle torque. In addition, because the foot inertia is small [14] compared to the impedance components, only the angular and linear acceleration acting on the plane of motion were considered in the impedance model.

$$T_F^z[t] \triangleq K_{DP} q^z[t] + B_{DP} \dot{q}^z[t] + J_{DP} \ddot{q}^z[t] + \alpha_{rot} \dot{\omega}_F^z[t] + \alpha_{linx} \ddot{p}_F^x[t] + \alpha_{liny} \ddot{p}_F^y[t]$$
$$(2.13)$$

$$T_F^x[t] \triangleq K_{IE} q^x[t] + B_{IE} \dot{q}^x[t] + J_{IE} \ddot{q}^x[t] + \beta_{rot} \dot{\omega}_F^z[t] + \beta_{linz} \ddot{p}_F^z[t] + \beta_{liny} \ddot{p}_F^y[t]$$
$$(2.14)$$

where K, B, J, and T_0, are the stiffness, damping, inertia, and mean torque, respectively. The operators superscript $\star^x$ and $\star^z$ select the x and z scalar components from the $\mathbb{R}^3$ vectors.

Signal processing. The impedance parameters were estimated for each subject, at each pose, combining samples from multiple perturbations (0.8 s window, starting on the onset of perturbation), and solving via least square regression. To reduce the effects of low-frequency active ankle torque and high-frequency noise in the estimation, a band-pass filter (3–35 Hz, 5th Order Butterworth) was applied on T_F, q, $\dot{q}$, $\ddot{q}$, $\dot{\omega}_F$, and $\ddot{p}_F$ (Fig. 2.7C–D). All the derivatives were numerically calculated via Savitzky-Golay filter (5th order polynomial in a 15-samples window).

Outlier removal. Finally, to account for the modulation of impedance due to sudden muscle contractions [2], samples with an absolute residual greater than 2.5 times the standard deviation of the residual were discarded as outliers; then the regression was recalculated. This process was repeated until there were no new outliers (11% of samples were discarded, and all impedance models had at least 65% inliers).

3 Results and discussion

3.1 Range of ankle torque and angle

The average COP position (in the anterior-posterior direction) was 28.1 ± 1.6, 40.7 ± 1.2, 52.7 ± 2.0, and 64.9 ± 2.9% of the foot length, for FF, MS, MS+, and TS, respectively. Because the COP position and ankle torque are proportional as

$$COP_x = \left(T_P^z / F_P^y - d_{P.heel} \right) / L_{foot} \qquad (2.15)$$

where $d_{P.heel}$ and L_{foot} are the foot length and distance from heel to the center of FP, the torque also increased monotonically (0.06 ± 0.04, 0.25 ± 0.03, 0.40 ± 0.05, and 0.60 ± 0.06 Nm/kg for FF, MS, MS+, and TS, respectively), and plantar-flexed in all poses. However, the ankle angle was plantar-flexed at FF (-11.9° ± 3.4°), stayed at a neutral angle for MS (0.5° ± 3.0°) and MS+ (1.8° ± 3.7°), and dorsiflexed at TS (10.7° ± 2.8°). The variance of mean values across subjects is explained by physiological differences and postural sway.

3.2 Dynamics of the experimental apparatus

The FP inertia parameters were estimated as $[35.0 \mp 1.9, 131.7 \mp 18.8, 119.6 \mp 2.5]^T$ g.m^2, $[0.1 \mp 1.2, 1,9 \mp 0.4, 0.4 \mp 1.2]^T$ g.m^2, and $[4.8 \mp 1.3, -6.9 \mp 2.9, 2.3 \mp 1.3]^T$ mm for the moment of inertia, the product of inertia, and the center of mass vector, respectively. The variance accounted for (VAF) of the torque and force reconstruction [(Eq. (2.1) and (2.2)] was $[95.7 \pm 1.3, 85.1 \pm 12.9, 98.0 \pm 0.8]^T$ and $[95.3 \pm 2.7, 85.4 \pm 4.3, 97.5 \pm 1.3]^T$, respectively. The mass parameter (mass of the components above the force sensor) was reported by the manufacturer as 4.64 kg and was fixed during the estimation.

The FP moment of inertia and product of inertia was similar to a box of equivalent size and mass. This box would have a moment of inertia of $[34.8, 131.2, 97.0]^T$ g.m^2 and 0.0 g.m^2 product of inertia. The small estimated product of inertia (consistent with the symmetric shape of the body) indicates that the mass inside the FP case is well distributed. The large variance of J_p^y estimated was expected because the vibrating platform cannot move in this axis of rotation; thus, the signal to noise ratio (SNR) around this axis is small. Similarly, the estimated center of mass vector has a high standard deviation, indicating a high uncertainty. Possibly, this certainty of this parameter could be higher if the plate underwent pure translation, rather than a constrained translation to the rotation (Fig. 2.8) (Table 2.2).

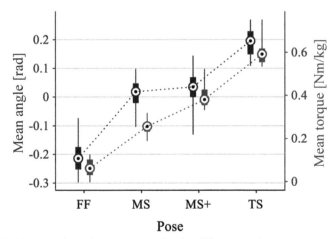

Figure 2.8 Mean angle and mean torque for the different static poses.

Table 2.2 Estimates of the FP inertia parameters.

	J_p [g.m^2]			P_p [g.m^2]				r_p [mm]	
Min	31.5	108.2	115.4	−2.2	1.2	−1.4	2.2	−10.0	−0.6
Max	37.0	149.6	122.7	1.3	2.4	2.0	6.7	−1.8	3.7
Mean	35.0	131.7	119.6	0.1	1.9	0.4	4.8	−6.9	2.3
Std	1.9	18.8	2.5	1.2	0.4	1.2	1.3	2.9	1.3

3.3 Mechanical impedance of the human ankle

The stiffness in the DP axis showed a strong correlation with the ankle torque, also increasing monotonically from 80.4 ± 23.4 Nm/rad in FF to 373.0 ± 94.4 Nm/rad in TS (Fig. 2.9). The other parameters did not show strong trend, averaging 75.0 ± 24.7 Nm/rad, 0.21 ± 0.50 Nms/rad, 0.09 ± 0.30 Nms/rad, 4.89 ± 12.08 × 10^{-3} kg.m^2, and 12.58 ± 6.27 × 10^{-3} kg. m^2, for IE stiffness, DP damping, IE damping, DP inertia, and IE inertia, respectively. A detailed description of the impedance estimates is shown in Table 2.3.

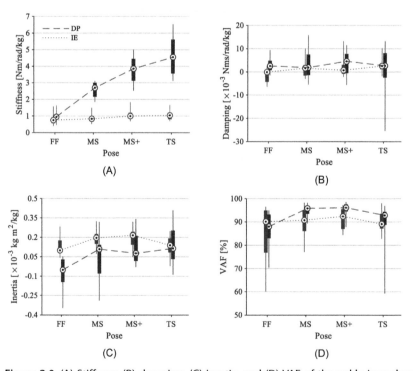

(A)

(B)

(C)

(D)

Figure 2.9 (A) Stiffness, (B) damping, (C) inertia, and (D) VAF of the ankle impedance model for the different static poses.

Table 2.3 Estimated impedance parameters group in different static gait poses.

		FF	MS	MS+	TS
K_{DP}	[Nm/rad]	79.5 ± 27.5	202.8 ± 40.1	300.7 ± 66.1	361.2 ± 81.4
K_{IE}	[Nm/rad]	63.2 ± 27.6	69.2 ± 23.6	84.3 ± 24.9	82.5 ± 21.2
B_{DP}	[× 10^{-3} Nms/rad]	248.2 ± 291.5	275.7 ± 529.8	290.0 ± 420.5	165.0 ± 723.1
B_{IE}	[× 10^{-3} Nms/rad]	-138.7 ± 247.4	112.7 ± 276.6	143.4 ± 325.9	208.7 ± 236.6
J_{DP}	[× 10^{-3} kg.m²]	-5.1 ± 9.2	4.1 ± 12.3	8.5 ± 8.1	12.1 ± 11.6
J_{IE}	[× 10^{-3} kg.m²]	10.0 ± 5.8	15.1 ± 5.4	14.7 ± 6.4	10.5 ± 6.2
VAF_{DP}	[%]	86.6 ± 7.6	4.8 ± 2.6	95.6 ± 2.8	89.7 ± 9.4
VAF_{IE}	[%]	85.7 ± 11.1	'89.9 ± 6.0	91.4 ± 4.4	90.3 ± 4.8

The average and standard deviation were calculated across subject estimates.

VAF for FF and TS were the lowest. Possibly because it was more challenging to maintain a static pose in these cases (loss of balance happened more often). The effort of regaining balance requires a response in the ankle torque, which might modulate the impedance [17], thus breaking the model assumption of constant impedance for that pose.

Negative damping and inertia might have been caused by large variance of the estimates since this parameter was small, close to zero. However, these parameters might be in fact negative because they might have a significant contribution from the reflex dynamics and complex muscle mechanics. The reflex dynamics are likely to affect these results because a large time window was used for the system identification (0.8 s from the onset of ground perturbations).

A repeated-measures analysis of variance (ANOVA) test was performed for the stiffness, damping, and inertia. A significant difference among poses was found on DP stiffness ($p < 0.001$), IE stiffness ($p = 0.006$), IE damping ($p < 0.001$), DP inertia ($p < 0.001$), and IE inertia ($p = 0.009$), but not for DP damping. A Tukey-Kramer test was used for post-hoc analysis to determine which pairs of static gait poses are significantly different, resulting in: all poses were significantly different from each other ($p < 0.01$) for DP stiffness; MS-MS+ ($p = 0.045$) for IE stiffness; FF-MS, FF-MS+, and FF-TS ($p < 0.01$) for IE damping; FF-MS+ and FF-TS ($p < 0.05$) for DP inertia; and no particular pair-wise group difference was for IE inertia.

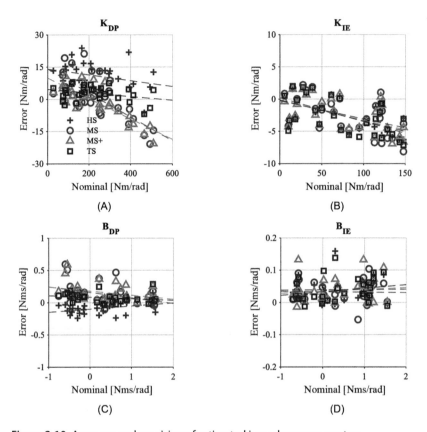

Figure 2.10 Accuracy and precision of estimated impedance parameters.

3.4 Validation using the simulation

The comparison between estimate error and reference stiffness and damping in IE and DP directions is presented in Fig. 2.10. Included in the plot are the estimates of the 30 simulated subjects at the 4 static gait poses totaling $30 \times 4 = 120$ points. Ideally, these points should all lie on the $y = 0$ line. However, the effects of the measurement noise and artifacts from numerical operations might spread or shift the points from this line.

The parameters estimated accurately with a normalized mean square error (NMSE) about 99%, except for the damping in DP direction (Table 2.4). B_{DP} was less accurate than B_{IE}, possibly because its relative contribution to the torque is smaller (comparing to DP). In other words, an angle noise affects more a damping estimate if the stiffness is large because a larger torque variance must be explained. This is the case for the DP

Table 2.4 NMSE of predicted simulated impedance parameters.

Pose	K_{DP}	K_{IE}	B_{DP}	B_{IE}
HS	99.1	99.4	97.2	99.5
MS	99.4	99.3	91.9	99.6
MS+	99.6	99.4	91.8	99.5
TS	99.9	99.3	97.8	99.5

Table 2.5 Bias errors of simulated ankle impedance coefficients separated by the (a) intercept and (b) slope of the residual curve.

	(a)					(b)			
Pose	K_{DP}	K_{IE}	B_{DP}	B_{IE}	Pose	K_{DP}	K_{IE}	B_{DP}	B_{IE}
HS	14.0*	−0.3	−0.10*	0.04*	HS	−0.014	−0.028*	0.048	0.001
MS	14.1*	0.2	0.16*	0.03*	MS	−0.056*	−0.037*	−0.085	−0.001
MS+	9.9*	−0.2	0.18*	0.03*	MS+	−0.048*	−0.031*	−0.063	0.006
TS	4.6*	−0.1	0.08*	0.03*	TS	−0.008	−0.033*	−0.024	0.011

* Represents coefficients significantly different than zero ($p < 0.01$, T-test).

direction, where the stiffness is 4 times greater while the damping is the same value.

The bias error verified in the nominal versus error curve (Fig. 2.10) was composed by two factors: a constant offset (intercept of the curve) and a growing error proportional to the nominal parameter (slope). The bias error skewed the estimation, in the worst-case scenario, by −19.3 Nm/rad, −5.3 Nm/rad, 0.24 Nms/rad, and 0.05 Nms/rad for K_{DP}, K_{IE}, B_{DP}, and B_{IE}, respectively. However, this represents only −3.2%, −3.5%, 8.1%, and 1.3% error when normalizing by the corresponding impedance range. The parameters of the residual line fit are shown in Table 2.5.

The VAF of the reconstructed torque (equivalent to Fig. 2.9D) was 97.9 ± 4.4%, without much variation between poses or DOF. Because the VAF from the human experiments were lower (90.9 ± 7.0%), this indicates the human ankle impedance is more complex than modeled in this study.

4 Conclusions

This study described a method to estimate the mechanical impedance in the dorsi–plantar and inversion–eversion degrees of freedom of the human ankle. This method was applied to 15 male subjects that reported never having a severe ankle injury or any biomechanical or neurological disorders. In addition, the same method was applied to 30 artificially created simulations to assess the accuracy of the impedance parameters.

The resulting mechanical impedance was similar to other non-functional experiments. Both stiffness and damping showed comparable ranges to supine experiments [2]. In addition, the stiffness showed an increasing trend with the increase of mean torque and dorsi ankle angle, as observed in the same study.

The mechanical impedance also showed comparable results to time-varying results from walking experiments. The stiffness in DP increased monotonically across the static gait poses (FF, MD, MD+, and TS) while the other parameters remained relatively constant. The increasing trend of DP stiffness was similar to results using ground perturbations, developed by Rouse [9], although the DP damping of their study was substantially larger. The DP damping was comparable to the swing phase, as studied by Lee [10]. Due to the similarity of the results to the time-varying impedance during gait, it opens the possibility to reconstruct the impedance as a function of the ankle angle and COP.

References

[1] P.L. Weiss, R.E. Kearney, I.W. Hunter, Position dependence of ankle joint dynamics—I. Passive mechanics, J. Biomech. 19 (9) (1986) 727–735.

[2] P.L. Weiss, R.E. Kearney, I.W. Hunter, Position dependence of ankle joint dynamics—II. Active mechanics, J. Biomech. 19 (9) (1986) 737–751.

[3] H. Lee, H.I. Krebs, N. Hogan, Multivariable dynamic ankle mechanical impedance with active muscles, IEEE Trans. Neural Syst. Rehabil. Eng. 22 (5) (2014) 971–981.

[4] H. Lee, P. Ho, M. Rastgaar, H.I. Krebs, N. Hogan, Multivariable static ankle mechanical impedance with active muscles, IEEE Trans. Neural Syst. Rehabil. Eng. 22 (1) (2014) 44–52.

[5] E.S. Tehrani, K. Jalaleddini, R.E. Kearney, Ankle joint intrinsic dynamics is more complex than a mass-spring-damper model, IEEE Trans. Neural Syst. Rehabil. Eng. 25 (9) (2017) 1568–1580.

[6] P. Amiri, R.E. Kearney, Ankle intrinsic stiffness changes with postural sway, J. Biomech. 85 (2019) 50–58.

[7] T.E. Sakanaka, J. Gill, M.D. Lakie, R.F. Reynolds, Intrinsic ankle stiffness during standing increases with ankle torque and passive stretch of the Achilles tendon, PloS One 13 (3) (2018) e0193850.

[8] E.M. Ficanha, G.A. Ribeiro, L. Knop, M. Rastgaar, Time-varying human ankle impedance in the sagittal and frontal planes during stance phase of walking, in: IEEE International Conference on Robotics and Automation (ICRA), 2017, pp. 6658–6664.

[9] E.J. Rouse, L.J. Hargrove, E.J. Perreault, T.A. Kuiken, Estimation of human ankle impedance during the stance phase of walking, IEEE Trans. Neural Syst. Rehabil. Eng. 22 (4) (2014) 870–878.

[10] H. Lee, N. Hogan, Time-varying ankle mechanical impedance during human locomotion, IEEE Trans. Neural Syst. Rehabil. Eng. 23 (5) (2015) 755–764.

[11] A.L. Shorter, E.J. Rouse, Mechanical impedance of the ankle during the terminal stance phase of walking, IEEE Trans. Neural Syst. Rehabil. Eng. 26 (1) (2018) 135–143.

[12] E.M. Ficanha, G.A. Ribeiro, L. Knop, M. Rastgaar, Time-varying impedance of the human ankle in the sagittal and frontal planes during straight walk and turning steps, in: International Conference on Rehabilitation Robotics (ICORR), 2017 pp. 1413–1418.

[13] E.M. Ficanha, G.A. Ribeiro, M. Rastgaar, Design and evaluation of a 2-DOF instrumented platform for estimation of the ankle mechanical impedance in the sagittal and frontal planes, IEEEASME Trans. Mechatron. 21 (5) (2016) 2531–2542.

[14] R.F. Chandler, C.E. Clauser, J.T. McConville, H.M. Reynolds, J.W. Young, Investigation of inertial properties of the human body, Air Force Aerospace Medical Research Lab Wright-Patterson AFB OH, 1975.

[15] Abraham, M.J.E. Savitzky, Golay, Smoothing and differentiation of data by simplified least squares procedures, Anal. Chem. 36 (8) (1964) 1627–1639.

[16] R.H. Byrd, J.C. Gilbert, J. Nocedal, A trust region method based on interior point techniques for nonlinear programming, Math. Program. 89 (1) (2000) 149–185.

[17] I.W. Hunter, R.E. Kearney, Dynamics of human ankle stiffness: variation with mean ankle torque, J. Biomech. 15 (10) (1982) 747–752.

CHAPTER 3

Task-dependent modulation of multi-dimensional human ankle stiffness

Hyunglae Lee

School for Engineering of Matter, Transport and Energy, Arizona State University, Tempe, AZ, United States

1 Introduction

1.1 Importance of human ankle stiffness

The human ankle plays one of the most important roles in lower extremity functions [1]. It contributes to not only anteroposterior and mediolateral stability during posture maintenance, but also propulsion, shock absorption, and lower-limb joint coordination during locomotion [2–5]. Proper control of the ankle allows accommodation of environmental changes and enables seamless physical interaction during daily motor tasks [6]. On the other hand, failure to properly control the ankle significantly impacts the overall quality of motor function, by making it more challenging and fatiguing as well as increasing the risk of unintentional falls.

Thus, for any robots physically interacting with the human lower extremities (e.g., lower limb exoskeleton robot, powered ankle-foot orthoses, and active leg prostheses), adequate control of the robotic ankle joint is critical to achieve coupled stability of the human–robot system and realize seamless physical human–robot interaction [8]. In order to aid in the design and control of lower extremity robots, it is important to first understand human ankle mechanics during various motor tasks.

1.2 Quasi-stiffness of the human ankle

In an effort to understand human ankle mechanics, the relationship between joint angle and torque at the ankle joint has been widely studied. Traditional motion analysis techniques using a motion capture system and a force plate have been widely used over the past few decades to investigate quasi-stiffness (the derivative of the torque-angle relationship with respect to angle) of the ankle during postural balance [4,10,11] and locomotion [12,13].

Powered Prostheses
http://dx.doi.org/10.1016/B978-0-12-817450-0.00003-1

41

However, it should be noted that quasi-stiffness and stiffness (position-dependent element that stores and releases energy [14]) are distinct concepts. Except when the joint is constrained to be passive, they can be significantly different during functional tasks that require significant muscle activation [15]. This distinction becomes even more important in the control of a robotic ankle joint because it can actively change not only joint stiffness but also equilibrium positions, providing a numerous combination of quasi-stiffness and stiffness [15]. Quantification of quasi-stiffness of the human ankle does not provide sufficient information for the impedance control of the robotic ankle joint, and thus quantification of human ankle stiffness is necessary.

1.3 Quantification of single-dimensional human ankle stiffness

In order to quantify joint stiffness without knowledge of the equilibrium positions, perturbations, that is, external energy inputs to excite the system, are required. Simple devices consisting of a servo-controlled motor and a cast supporting the leg have been widely utilized, in combination with system identification techniques, to estimate ankle stiffness in seated or supine posture [16–21]. Platform devices rotating the footplate have been also used to estimate ankle stiffness during upright standing [10,11,22–24]. These studies demonstrated that ankle stiffness is highly dependent on the level of ankle torque, muscle co-contraction, and ankle position. This finding is consistent with a well-known analytical model of human joint mechanics explaining that human joint stiffness is determined by the summation of muscle generated stiffness and kinematic stiffness originated from nonlinear musculotendon kinematics [25–27].

However, most of the previous studies characterizing ankle stiffness have focused on a single degree-of-freedom (DOF) of the ankle, in particular, dorsiflexion-plantarflexion (DP) in the sagittal plane. Only a few studies reported quantification of ankle stiffness in inversion-eversion (IE) in the frontal plane, but still limited to a single DOF of the ankle [22,23,28,29].

2 Task-dependent modulation of multi-dimensional human ankle stiffness

2.1 Importance of multi-dimensional human ankle stiffness

The human ankle is a complicated joint involving multiple bones, ligaments, tendons, and muscles, and neural commands from the brain and spinal reflex feedback may further complicate actions of the ankle joint in multiple DOFs. Single DOF ankle movements are unusual during normal lower-extremity functions, which include postural balance and walking [6]. Thus, characterization of multi-dimensional ankle stiffness will not only promise deeper understanding of its roles in motor control and function, not achievable from single DOF studies, but also provide insights for the design and control of multi-DOF robotic ankle joint of lower-extremity robots.

2.2 Quantification of 2-dimensional human ankle stiffness during seated tasks

As a baseline for understanding ankle stiffness modulation during functional tasks, we first quantified 2-dimensional ankle stiffness in a seated position (Fig. 3.1A). In this study, we utilized a wearable ankle robot (Anklebot; Bionik Laboratories Corp., Canada), capable of providing actively controllable torques in 2 DOF of the ankle, spanning the sagittal and frontal planes (Fig. 3.1B). The 3rd DOF (axial rotation) is passive with extremely low friction, thereby avoiding imposing any inadvertent kinematic constraints on the motion of the ankle. The robot is highly backdrivable with minimal internal frictional forces and low inertia. During experiments, subjects wore a knee brace, which was connected to the side plate of the chair. The brace did not limit the subjects' ability to flex or extend their knee but helped subjects relax their knee throughout the experiment. The body of the ankle robot was mounted to the knee brace and two actuators were connected to the U-shaped bracket of the shoe through ball joints (Fig. 3.1C).

In this study, we quantified 2-dimensional ankle stiffness over a wide range of muscle activation levels; relaxed, 10%−30% of maximum voluntary

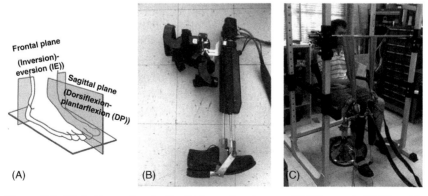

Figure 3.1 (A) Ankle motion in the 2-dimensional space (IE-DP space). (B) Wearable ankle robot, (C) Experimental Setup in the seated posture.

contraction (MVC) levels of tibialis anterior (TA; ankle dorsiflexor), and 10%−30% MVC levels of soleus (SOL; ankle plantarflexor), with increments of 5% MVC. During measurements, a visual feedback display showing current and target activation levels was provided to subjects. Subjects were first instructed to activate a specific muscle and maintain it at the target level. When the activation level reached the target level, the robot applied mild random torque perturbations (band-limited white noise with a spectrum flat up to 100 Hz) to the ankle for 40 s for each measurement.

Linear time-invariant multi-input multi-output system identification based on spectral analysis [30] was applied to characterize the directional variation of ankle stiffness in the 2-dimensional space (IE-DP space formed by IE and DP axes), the relationship between muscle activation and the corresponding ankle stiffness, and the amount of coupling between the 2 DOFs [31,32].

Briefly, the system identification method first quantified frequency response of ankle dynamics. In estimating ankle dynamics, the contribution of actuator dynamics was properly compensated. Next, stiffness component of ankle dynamics was calculated by averaging the magnitude frequency response in a low frequency region below 5 Hz, where stiffness dominates the dynamic response. By repeating this identification process for different movement directions in the IE-DP space, 2-dimensional ankle stiffness was estimated. The amount of coupling between the 2 DOFs was quantified from a 2-by-2 partial coherence matrix, which indicates linear dependency between each input and output after removing the effects of other inputs [30]. Detailed descriptions of the methods and validation are provided in Ref. [31].

This study for the first time provided several important ankle characteristics in 2 DOFs unavailable from previous single DOF studies. First, ankle stiffness was highly direction dependent, being greater in the sagittal plane than in the frontal plane at all muscle activation conditions (Fig. 3.2A; Table 3.1). The shape of directional variation was consistent over the range of muscle activation levels tested in this study; weaker stiffness in IE than DP resulted in a characteristic "peanut" shape. Activating muscles significantly increased ankle stiffness in all directions in the IE − DP space, but it increased more in DP than in IE, accentuating the "peanut" shape, pinched in the IE direction. Although the amount of stiffness increase was greater in DP than IE, the ratio of DP stiffness to IE stiffness did not change significantly with muscle activation level. In addition, the ratio of active stiffness to passive (maximally-relaxed) stiffness was calculated; the ratio in IE was sub-

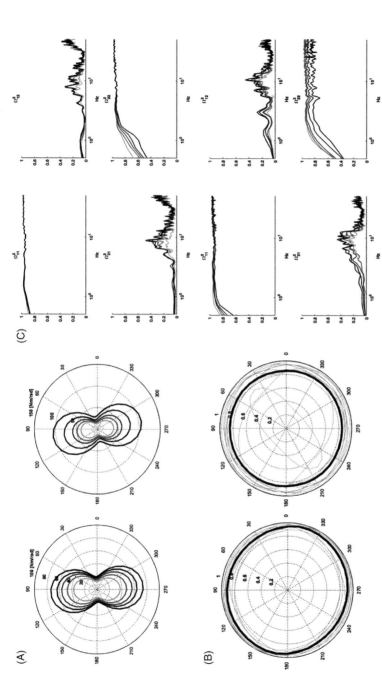

Figure 3.2 (A) Directional variation of ankle stiffness. The angle denotes movement direction in the 2D space and the radius represents the magnitude of impedance in the movement direction. Right foot eversion, dorsiflexion, inversion and plantarflexion, correspond to $0°$, $90°$, $180°$ and $270°$, respectively. *(Left)*: TA active study, *(Right)*: SOL active study. Results at target level 10%, 15%, 20%, 25%, and 30% MVC are presented with colors from lighter to darker. The means of all subjects are presented. (B) Linearity between muscle activation level and ankle stiffness in IE − DP space. The correlation coefficient (R^2) for each movement direction is presented in polar coordinates. Each thin *grey line* represents the result of an individual subject, and the thick *black line* represents the mean of all subjects. *(Left)*: TA active study, *(Right)*: SOL active study. (C) Partial coherences in IE-DP coordinates. *(Top)*: TA active study. *(Bottom)*: SOL active study.

Table. 3.1 Directional variation of 2-dimensional ankle stiffness.

The ratio of DP stiffness to IE stiffness

Study Target level	TA active study	SOL active study
10% MVC	3.62 (0.16)	3.56 (0.40)
15% MVC	3.75 (0.17)	3.61 (0.44)
20% MVC	3.78 (0.18)	3.58 (0.51)
25% MVC	3.79 (0.17)	3.54 (0.51)
30% MVC	3.78 (0.19)	3.35 (0.45)

The ratio of total (passive + active) stiffness to passive stiffness

Study	TA active study		SOL active study	
Direction Target level	IE	DP	IE	DP
10% MVC	1.60 (0.13)	3.15 (0.25)	1.87 (0.10)	3.41 (0.50)
15% MVC	2.03 (0.19)	4.11 (0.37)	2.46 (0.17)	4.23 (0.58)
20% MVC	2.32 (0.21)	4.84 (0.49)	3.11 (0.22)	5.22 (0.75)
25% MVC	2.65 (0.22)	5.49 (0.50)	3.73 (0.22)	6.66 (1.33)
30% MVC	2.91 (0.24)	6.05 (0.57)	4.98 (0.25)	8.94 (1.83)

The mean and standard error (SE, in parentheses) across all 10 subjects are presented in the table.

stantially lower than in DP for each activation levels, implying that the relative contribution of passive stiffness to total stiffness (passive stiffness + active stiffness) was significantly higher in IE than DP (Table 3.1). Next, the relationship between muscle activation and ankle impedance was highly linear in all movement directions in the 2-dimensional space [32]. The correlation coefficient (R^2) between the level of muscle activation and the corresponding ankle stiffness was calculated and presented in polar coordinates (Fig. 3.2B). In the TA active study, R^2 values were very high: 0.94, 0.95, and 0.94 for IE, DP, and all movement directions, respectively, when averaged across all subjects. In the SOL active study, R^2 values were somewhat lower though still high: 0.87, 0.87, and 0.87 for IE, DP, and all movement directions, respectively. Finally, the coupling between DOFs was small even with significant muscle activation. This was evidenced by low off-diagonal partial coherences (< 0.2) at most frequencies, except in a region around 10 Hz (Fig. 3.2C). In this region, partial coherences were higher than those in the relaxed study [31], and increased with muscle activation. However, even the highest coherence value was still low (~ 0.4).

In summary, despite the internal complexities of the ankle, its 2–dimensional ankle stiffness was remarkably simple, at least for our young healthy subjects and in the context of this experiment; the coupling between IE and DP was small so that IE and DP stiffness can be treated independently; and the relationship between muscle activation and ankle stiffness in the 2–dimensional space was highly linear. Considering that multiple ankle muscles acting in both the sagittal and frontal planes affect ankle behavior in both DOFs and intra- and inter-muscular reflex feedback may also contribute to the modulation of ankle impedance [33], it is surprising that multi–dimensional ankle stiffness is externally simple even with considerable muscle activation.

2.3 Quantification of 2-dimensional human ankle stiffness during walking tasks

While many previous studies have characterized ankle stiffness in static task conditions including our previous studies on the 2-dimensional human ankle stiffness during seated tasks [27,31,32,34], little is known about how ankle stiffness is modulated during dynamic tasks, mainly due to the lack of proper quantification methods or tools. To address this limitation, we proposed a new approach of integrating the wearable ankle robot and linear time-varying system identification methods and quantified 2-dimensional ankle stiffness as it varies with time during walking.

In this study, we quantified 2-dimensional ankle stiffness simultaneously from pre-swing through swing to early stance. This included heel–strike (HS) and toe–off (TO), two key events in the transition from swing to stance or vice versa (Fig. 3.3A). Subjects wore the wearable ankle robot

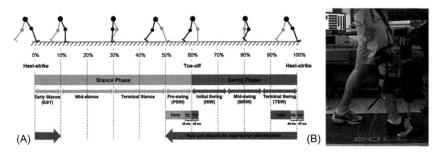

Figure 3.3 (A) Sub-gait phases for ankle stiffness quantification. Stiffness was identified in 5 sub-gait phases: pre-swing (PSW), initial swing (ISW), mid-swing (MSW), terminal swing (TSW), and early stance (EST). Time periods for closer examination around heel-strike (HS) and toe-off (TO) are illustrated. (B) Experimental setup for the walking task.

and were instructed to walk on a treadmill at their preferred walking speed (Fig. 3.3B). After 1 min of walking without actuation of the robot, mild random torque perturbations were applied to the ankle for the next 13 min; independent band-limited white noise with stop frequency 100 Hz was applied to each actuator to produce random torque perturbations at the ankle joint in 2 DOFs. The magnitude of perturbation was selected strong enough to perturb the ankle from pre-swing to early stance, but low enough not to disturb normal walking.

Ankle stiffness was estimated based on ensemble-based linear time-varying system identification. This is an effective and robust system identification technique when repetitive and periodic data are available. In addition, it requires no a priori knowledge of the structure of the system to be identified and can capture fast time-varying behavior of the system [35]. In this study, more than 500 realizations for ensemble sets were generated from human walking, where each realization was defined by a gait cycle that begins with HS of one foot and ends with another HS of the same foot. Input and output ensemble sets contain torque perturbations and the resulting ankle displacements, respectively. By applying correlation-based system identification approach [35] to the ensemble sets, measured dynamics were estimated along the time axis (at 2 ms intervals) in the form of finite impulse response functions (IRF). The IRF estimates were further approximated by second order models consisting of stiffness, damping, and foot inertia. Finally, ankle stiffness, damping, and foot inertia were obtained by compensating actuator dynamics. While ankle stiffness was estimated every 2 ms, representative ankle parameters were calculated at five sub-gait phases, ranging from 0%−10% (early stance: EST), 50%−60% (pre-swing phase: PSW), 60%−73% (initial swing phase: ISW), 74%−86% (mid-swing phase: MSW), and 87%−100% (terminal swing phase: TSW) of the gait cycle by averaging estimated parameters within each sub-gait phase into single values. Detailed descriptions of the methods and validation are provided in Ref. [36].

Ankle stiffness exhibited significant time-varying behaviors in both DOFs during walking. It significantly decreased at the end of stance before TO, remain relatively constant across the swing phase, and increased around HS (Fig. 3.4A). Closer investigation around HS revealed that 2-dimensional ankle stiffness started to increase in terminal swing before HS. These "pre-tuning" behaviors of the ankle may be essential to maintain natural and stable human walking. For example, lowering ankle stiffness before TO may assist dorsiflexion of the foot to provide toe clearance in initial swing.

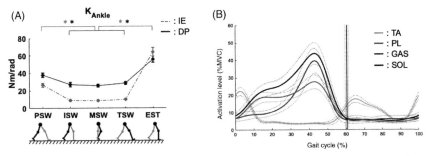

Figure 3.4 (A) Time-varying ankle stiffness in gait sub-phases. Representative values for 5 gait sub-phases were obtained by averaging results within each gait sub-phase. *Dashed* lines and *solid* lines denote results in IE and DP directions, respectively. Asterisks denote statistical difference ($p < 0.05$). The mean and mean ± 1 SE of all 13 subjects are illustrated as asterisks and bars, respectively. (B) Normalized EMG amplitude in one gait cycle. The mean *(solid lines)* and mean ± 1 SE *(dotted lines)* of all subjects are illustrated. The *vertical line* denotes TO mean *(solid line)* and mean ± 1 SE *(dotted lines)*.

Increased ankle stiffness before HS may prepare for "shock–absorption" in the loading response right after HS. The pattern of stiffness change was consistent with decreasing activation of plantarflexors from the end of terminal stance to initial swing and increasing activation of both dorsiflexor and plantarflexors, that is, co-contraction, around HS (Fig. 3.4B). In addition, ankle stiffness was greater in DP than IE across all phases, except right after HS. Ankle stiffness in IE after HS was significantly higher than what we observed during seated task with substantial muscle activation. This significant increase in ankle stiffness cannot be explained by muscle co-contraction alone. One possible explanation is that the ankle joint "locks" in the frontal plane right after HS. It has been reported that the subtalar joint, which is responsible for IE movements, locks with eversion when HS occurs [37]. Furthermore, it has been reported that the mortise of the ankle, the space formed by the top of the talus and the lower ends of the tibia and the fibula, locks in the frontal plane at the moment of HS [38]. Whatever its origin, the substantial increase of frontal plane ankle impedance serves to improve lateral stability from the moment of HS throughout the loading response. The patterns for 2-dimensional ankle stiffness explained earlier were consistent across all 13 subjects who participated in this study.

One limitation of this study is that reliable quantification cannot be made when the loading at the ankle joint is substantial, for example, during mid- or terminal stance in walking. To address this limitation, three recent studies have developed robotic platforms that can provide perturbations from the ground [39–41]. Rouse and colleagues developed a 1-DOF

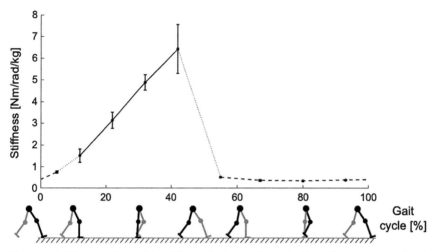

Figure 3.5 *Time-varying ankle stiffness during walking.* Ankle stiffness is normalized by bodyweight and summarized at nine timing points. The stance and swing phases account for approximately 60% and 40% of the gait cycle. *Solid* and *dashed* lines denote results from [7] and [9], respectively. Asterisks and error bars denote the mean and SE, respectively.

position–controlled robotic platform and used this platform to quantify ankle stiffness from late loading response to terminal stance phase [7,42]. Integration of these two studies with our study completed a trajectory of ankle stiffness modulation in the sagittal plane throughout the gait cycle (Fig. 3.5) [5]. Ficanha and colleagues developed a 2-DOFs torque-controlled vibrating platform and used this platform to quantify 2-dimensional ankle stiffness during stance [43]. Recently, we also developed a similar 2-DOFs robotic platform, which can simulate realistic physical environments. Details of this platform are described in Ref. [41] and a brief description is provided in the following section.

2.4 Quantification of 2-dimensional human ankle stiffness during standing tasks

Mounting evidence demonstrates that human ankle stiffness plays an integral role in the maintenance of postural balance during standing, known as the ankle strategy [4,6,44,45]. Since ankle stiffness can be modulated with various muscle activation patterns depending on motor tasks and physical environments that we interact with, it is important to understand the activation-dependent modulation of ankle stiffness. We used the multi-axis robotic platform to quantify 2-dimensional ankle stiffness when subjects performed standing tasks that involve a wide range of muscle activations including unilateral activation and co-contraction.

As a centerpiece of this study, we used the multi-axis robotic platform capable of providing fast position perturbations to the ankle joint in 2 DOFs, that is, DP and IE, and recording the corresponding torques at the ankle in 2 DOFs (Fig. 3.6A). A series of validation experiments demonstrated that the platform can provide rapid perturbations up to an angular velocity of 100°/s with an error less than 0.1° even under excessive loading [41]. The platform not only provided perturbations to the ankle but also measured the center of pressure (CoP), which was one important parameter to be controlled by human subjects throughout the experiment. The entire experimental setup consisted of the multi-axis robotic platform, a weight scale, a dual-axis goniometer, surface electromyography (EMG) sensors, a body weight support, and a visual feedback display (Fig. 3.6B). The weight scale was used to control the equal weight distribution between both legs. The dual-axis goniometer was used to measure ankle angles in 2 DOFs. Surface EMG sensors were also used to measure the muscle activity of four primary ankle muscles: TA, SOL, peroneus longus (PL), and medial gastrocnemius (GAS). To ensure safety of subjects, each subject wore a safety harness fastened to the body weight support system. The visual feedback display showed the

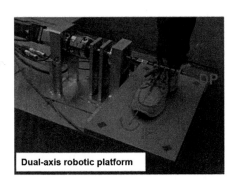

Dual-axis robotic platform

(A)

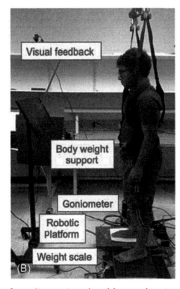

Visual feedback

Body weight support

Goniometer

Robotic Platform

Weight scale

(B)

Figure 3.6 (A) Dual-axis robotic platform to quantify 2-dimensional ankle mechanics during various lower extremity tasks. (B) Experimental setup for the standing study.

target and current levels of three parameters to be controlled by subjects: weight distribution between both legs, CoP displacements in the DP or IE direction, and TA muscle activation. While controlling the CoP alone modulated unilateral muscle activation, simultaneously controlling the CoP and TA activation level modulated muscle co-contraction. Thus, by varying the target CoP in the sagittal or frontal plane or changing the target TA activation level (CoP measured during quiet upright standing was used as a target in this condition), subjects could achieve a wide range of muscle activation in the control of postural balance during upright standing.

Subjects were instructed to maintain 1 of the following 12 experimental conditions: +10, +12.5, +15, +17.5 cm CoP in DP (sagittal plane), −1.5, −0.75, +0.75, +1.5 cm CoP in IE (frontal plane), and 0%, 10%, 20%, 30% maximum voluntary contraction (MVC) of TA. Controlling positive CoP in DP was intended to facilitate the activation of ankle plantarflexors. Similarly, positive and negative CoP in IE were designed to facilitate the activation of ankle inverters and everters, respectively. On the other hand, controlling 0 cm CoP with TA activation were intended to co-contract all ankle muscles. When the target conditions were satisfied, the robotic platform applied rapid 3 degrees ramp-and-hold perturbations to the ankle, with 75 ms duration. Ankle stiffness was calculated by fitting a 2nd order model, consisting of ankle stiffness, ankle damping, and foot inertia, to the measured ankle kinematics and torques. Linear regression was performed over an interval of 75 ms (the duration of the perturbation).

Two-dimensional ankle stiffness increased with increasing muscle activations. Increasing CoP in the DP direction resulted in a significant increase in DP ankle stiffness, with a minimum average stiffness of 237.2 Nm/rad for the 10 cm CoP condition (close to the CoP measured during quiet standing, i.e., 0% MVC of TA) and a maximum average stiffness of 467.8 Nm/rad for the 17.5 cm CoP condition where the activation of plantarlexors were the highest among the given experimental conditions (Fig. 3.7A). Co-contraction of ankle muscles also linearly increased DP ankle stiffness, with a minimum average stiffness of 235.3 Nm/rad at 0% MVC, and a maximum average stiffness of 489.0 Nm/rad at 30% MVC. Increasing CoP in the IE direction also resulted in a significant increase in IE ankle stiffness. However, negative CoP values (medial) yielded a significantly higher ankle stiffness than positive CoP (lateral), with a minimum difference in stiffness (between −0.75 cm and +0.75 cm CoP) of 22.65 Nm/rad and a maximum difference (between −1.5 cm and +1.5 cm CoP) of 37.96 Nm/rad (Fig. 3.7B). Co-contraction of ankle muscles also linearly increased IE ankle stiffness,

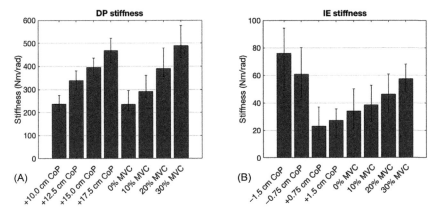

Figure 3.7 *Two-dimensional ankle stiffness during standing tasks with a range of muscle activation.* The CoP was controlled for the modulation of unilateral muscle activation and the TA muscle and the reference CoP (measured in quiet standing) were simultaneously controlled for the modulation of co-contraction of ankle muscles. (A) DP stiffness. (B) IE stiffness. The mean and standard deviation (errorbar) of 8 subjects are presented.

with a minimum average stiffness of 33.73 Nm/rad at 0% MVC, and a maximum average stiffness of 57.33 Nm/rad at 30% MVC.

In summary, 2-dimensional ankle stiffness significantly increased with unilateral muscle activation and co-contraction during standing tasks. In addition, DP stiffness was always greater than IE stiffness in all experimental conditions. While these findings were consistent with our previous findings, the magnitude of 2-dimensional ankle stiffness was greater than that observed during static seated tasks [27,32].

3 SUMMARY

This chapter has summarized our recent efforts to understand task-dependent modulation of 2-dimensional ankle stiffness. It should be noted that some contents in this chapter (texts, table, and figures) are excerpted from our previous publications with permission to re-use or reproduce them [9,32,41].

The use of multi-DOF robotic devices (the 2-DOFs wearable ankle robot and dual-axis robotic platform), not available in previous studies for the quantification of quasi-stiffness and single-dimensional ankle stiffness, enabled reliable quantification of 2-dimensional ankle stiffness (i.e., DP stiffness in the sagittal plane and IE stiffness in the frontal plane) during various tasks (seated, walking, and standing tasks) which involve a wide range of ankle muscle activation.

One important finding consistent across all tasks is that the human ankle is significantly stiffer in the sagittal plane than in the frontal plane. Within each motor task, muscle activation patterns were highly correlated with the modulation of 2-dimensional ankle stiffness. However, comparison of results across different tasks suggested that the loading condition at the ankle as well as muscle activation patterns is also an important factor determining the modulation of ankle stiffness. For example, ankle stiffness during standing tasks, where substantial loading was applied to the ankle joint, was significantly greater than that during unloaded seated tasks even for the comparable muscle activation level.

Active research has been conducted on the design and control of lower extremity robots such as powered ankle-foot orthoses, active leg prostheses, and exoskeletons that mimic intact human attributes, in particular human ankle stiffness [46–49]. To this end, it is essential to develop a reliable and robust model that estimates human ankle stiffness. Most of stiffness models in the current state-of-the-art studies cannot explain task-dependent modulation of ankle stiffness as well as its multi-dimensional characteristics, which might diminish the efficacy of the models. We have been tackling this problem and currently extending the quantification of multi-dimensional ankle stiffness to task situations that include a wide range of physical or mechanical environments, for example, soft and compliant grounds [50]. Ultimately, integration of findings observed from a wide range of task conditions will allow us to construct a universal ankle stiffness model that better explains task-dependent modulation of multi-dimensional ankle stiffness and provides more reliable information to robotic controllers.

References

[1] D.A. Winter, Biomechanics and Motor Control of Human Gait: Normal, Elderly and Pathological, Waterloo Biomechanics, second ed., Canada 1991.

[2] D.G. Robertson, D.A. Winter, Mechanical energy generation, absorption and transfer amongst segments during walking, (in eng), J. Biomech. 13 (10) (1980) 845–854.

[3] R.R. Neptune, S.A. Kautz, F.E. Zajac, Contributions of the individual ankle plantar flexors to support, forward progression and swing initiation during walking, J. Biomech. 34 (11) (2001) 1387–1398.

[4] D.A. Winter, A.E. Patla, S. Rietdyk, M.G. Ishac, Ankle muscle stiffness in the control of balance during quiet standing, J. Neurophysiol. 85 (6) (2001) 2630–2633.

[5] H. Lee, E. Rouse, H.I. Krebs, Summary of human ankle mechanical impedance during walking, IEEE J. Transl. Eng. Health Med. 4 (2016) 2100407.

[6] J. Perry, Gait Analysis: Normal and Pathologic Functions, Slack Inc., New Jersey, 1992.

[7] E.J. Rouse, L.J. Hargrove, E.J. Perreault, T.A. Kuiken, Estimation of human ankle impedance during the stance phase of walking, IEEE Trans. Neural Syst. Rehabil. Eng. 22 (4) (2014) 870–878.

[8] H. Lee, N. Hogan, Essential considerations for design and control of human-interactive robots, in: In Proceedings from IEEE International Conference on Robotics, Automation, (ICRA), 2016, Stockholm, pp. 3069–3074.

[9] H. Lee, N. Hogan, Time-varying ankle mechanical impedance during human locomotion, IEEE Trans. Neural Syst. Rehabil. Eng. 23 (5) (2015) 755–764.

[10] M. Casadio, P.G. Morasso, V. Sanguineti, Direct measurement of ankle stiffness during quiet standing: implications for control modelling and clinical application, Gait Posture 21 (4) (2005) 410–424.

[11] I.D. Loram, M. Lakie, Direct measurement of human ankle stiffness during quiet standing: the intrinsic mechanical stiffness is insufficient for stability, J. Physiol. 545 (3) (2002) 1041–1053.

[12] K. Shamaei, G.S. Sawicki, A.M. Dollar, Estimation of quasi-stiffness and propulsive work of the human ankle in the stance phase of walking, PLoS One 8 (3) (2013) e59935.

[13] A.H. Hansen, D.S. Childress, S.C. Miff, S.A. Gard, K.P. Mesplay, The human ankle during walking: implications for design of biomimetic ankle prostheses, J. Biomech. 37 (10) (2004) 1467–1474.

[14] M.L. Latash, V.M. Zatsiorsky, Joint stiffness—myth or reality, Human Movement Sci. 12 (6) (1993) 653–692.

[15] E.J. Rouse, R.D. Gregg, L.J. Hargrove, J.W. Sensinger, The difference between stiffness and quasi-stiffness in the context of biomechanical modeling, IEEE Trans. Biomed. Eng. 60 (2) (2013) 562–568.

[16] R.E. Kearney, R.B. Stein, L. Parameswaran, Identification of intrinsic and reflex contributions to human ankle stiffness dynamics, IEEE Trans. Biomed. Eng. 44 (6) (1997) 493–504.

[17] T. Sinkjaer, E. Toft, S. Andreassen, B.C. Hornemann, Muscle-stiffness in human ankle dorsiflexors—intrinsic and reflex components, J. Neurophysiol. 60 (3) (1988) 1110–1121.

[18] L.Q. Zhang, S.G. Chung, Y. Ren, L. Liu, E.J. Roth, W.Z. Rymer, Simultaneous characterizations of reflex and nonreflex dynamic and static changes in spastic hemiparesis, J. Neurophysiol. 110 (2) (2013) 418–430.

[19] I.W. Hunter, R.E. Kearney, Dynamics of human ankle stiffness - variation with mean ankle torque, J. Biomech. 15 (10) (1982) 747–752.

[20] R.E. Kearney, I.W. Hunter, Dynamics of human ankle stiffness—variation with displacement amplitude, J. Biomech. 15 (10) (1982) 753–756.

[21] R.E. Kearney, I.W. Hunter, System-identification of human joint dynamics, Crit. Rev. Biomed. Eng. 18 (1) (1990) 55–87.

[22] S.M. Zinder, K.P. Granata, D.A. Padua, B.M. Gansneder, Validity and reliability of a new in vivo ankle stiffness measurement device, J. Biomech. 40 (2) (2007) 463–467.

[23] T. Kobayashi, A.K.L. Leung, Y. Akazawa, M. Tanaka, S.W. Hutchins, Quantitative measurement of spastic ankle joint stiffness using a manual device: a preliminary study, J. Biomech. 43 (9) (2010) 1831–1834.

[24] J.R. Chagdes, S. Rietdyk, M.H. Jeffrey, N.Z. Howard, A. Raman, Dynamic stability of a human standing on a balance board, J. Biomech. 46 (15) (2013) 2593–2602.

[25] R. Shadmehr, M.A. Arbib, A mathematical-analysis of the force-stiffness characteristics of muscles in control of a single joint system, Biol. Cybernetics 66 (6) (1992) 463–477.

[26] N. Hogan, Mechanical impedance of single- and multi-articular system, in: J. Winters, S. Woo (Eds.) Multiple Muscle Systems: Biomechanics and Movement Organization, Springer-Verlag, New York, pp. 149–164, 1990.

[27] H. Lee, P. Ho, M. Rastgaar, H.I. Krebs, N. Hogan, Multivariable static ankle mechanical impedance with active muscles, IEEE Trans. Neural Syst. Rehabil. Eng. 22 (1) (2014) 44–52.

[28] A. Saripalli, S. Wilson, Dynamic ankle stability and ankle orientation, in: In Proceedings of the 7th symposium on footwear biomechanics, Cleveland, OH, 2005, pp. 1–2.

[29] J. Mizrahi, Y. Ramot, Z. Susak, The dynamics of the subtalar joint in sudden inversion of the foot, J. Biomech. Eng. 112 (1) (1990) 9–14.

[30] J. Bendat, A. Piersol, Random Data: Analysis and Measurement Process, 4th edition, Wiley, 2010.

[31] H. Lee, H.I. Krebs, N. Hogan, Multivariable dynamic ankle mechanical impedance with relaxed muscles, IEEE Trans. Neural Syst. Rehabil. Eng. 22 (6) (2014) 1104–1114.

[32] H. Lee, H.I. Krebs, N. Hogan, Multivariable dynamic ankle mechanical impedance with active muscles, IEEE Trans. Neural Syst. Rehabil. Eng. 22 (5) (2014) 971–981.

[33] A. Prochazka, Proprioceptive Feedback and Movement Regulation, Comprehensive Physiology, John Wiley & Sons, Inc., 2011, 1996.

[34] H. Lee, P. Ho, M.A. Rastgaar, H.I. Krebs, N. Hogan, Multivariable static ankle mechanical impedance with relaxed muscles, J. Biomech. 44 (10) (2011) 1901–1908.

[35] M. Lortie, R.E. Kearney, Identification of physiological systems: estimation of linear time-varying dynamics with non-white inputs and noisy outputs, Med. Biol. Eng. Comput. 39 (3) (2001) 381–390.

[36] H. Lee, N. Hogan, Time-varying ankle mechanical impedance during human locomotion, IEEE Trans. Neural Syst. Rehabil. Eng. 23 (5) (2015) 755–764.

[37] A.L. Logan, L.J. Rowe, The Foot and Ankle: Clinical Applications, Jones & Bartlett Learning, (1994).

[38] A. Arndt, P. Westblad, I. Winson, T. Hashimoto, A. Lundberg, Ankle and subtalar kinematics measured with intracortical pins during the stance phase of walking, Foot Ankle Int. 25 (5) (2004) 357–364.

[39] E.J. Rouse, L.J. Hargrove, E.J. Perreault, M.A. Peshkin, T.A. Kuiken, Development of a mechatronic platform and validation of methods for estimating ankle stiffness during the stance phase of walking, J. Biomech. Eng. Trans. ASME 135 (8) (2013) 81009.

[40] E.M. Ficanha, G. Ribeiro, M.A. Rastgaar, Design and evaluation of a 2-DOF instrumented platform for estimation of the ankle mechanical impedance in the sagittal and frontal planes, IEEE/ASME Trans. Mech. 21 (5) (2016) 2531–2542.

[41] V. Nalam and H. Lee, Development of a two-axis robotic platform for the characterization of two-dimensional ankle mechanics, IEEE/ASME Trans. Mech., 24 (2) (2019) 459–470.

[42] A.L. Shorter, E.J. Rouse, Mechanical impedance of the ankle during the terminal stance phase of walking, IEEE Trans. Neural Syst. Rehabil. Eng. 26 (1) (2018) 135–143.

[43] E. Ficanha, G. Ribeiro, L. Knop, M. Rastgaar, Estimation of the two degrees-of-freedom time-varying impedance of the human ankle, J. Med. Devices 12 (1) (2018) 1–5.

[44] D.A. Winter, Human balance and posture control during standing and walking, Gait Posture 3 (1995) 193–214.

[45] P. Gatev, S. Thomas, T. Kepple, M. Hallett, Feedforward ankle strategy of balance during quiet stance in adults, J. Physiol. 514 (Pt 3) (1999) 915–928.

[46] F. Sup, A. Bohara, M. Goldfarb, Design and control of a powered transfemoral prosthesis, Int. J. Robot. Res. 27 (2) (2008) 263–273.

[47] L.J. Hargrove, et al. Robotic leg control with EMG decoding in an amputee with nerve transfers, N Engl. J. Med. 369 (13) (2013) 1237–1242.

[48] S.K. Au, H.M. Herr, Powered ankle-foot prosthesis—the importance of series and parallel motor elasticity, IEEE Robot. Automat. Mag. 15 (3) (2008) 52–59.

[49] E.M. Ficanha, G.A. Ribeiro, H. Dallali, M. Rastgaar, Design and preliminary evaluation of a two DOFs cable-driven ankle-foot prosthesis with active dorsiflexion-plantarflexion and inversion-eversion, Front. Bioeng. Biotechnol. 4 (36) (2016).

[50] V. Nalam, H. Lee, Environment-dependent modulation of human ankle stiffness and its implication for the design of lower extremity robots, in: The 15th International Conference on Ubiquitous Robots (UR 2018), Hawaii, 2018.

CHAPTER 4

Kriging for prosthesis control

Neil Dhir

The Alan Turing Institute British Library, London

In this chapter we present the use of Bayesian nonparametric methods in the realm of powered prostheses control. More specifically we are interested in *change*; change in the driving functions of human locomotion behavior in response to endogenous and exogenous influences – see Fig. 4.1 for a high-level understanding of the influences involved.

Supervised learning within the realm of prosthesis control requires training examples. These building blocks or *tasks* need to be extracted from some set of observations, originating from some observation stream via some sensors. Those observations originate in the operational environment of the prosthesis and can be measured by way of time-series observations of the environment (which includes the dynamics of the prosthesis as well). Performing inference on those observations, to extract the underlying behavior inherent from user-environment interaction, is not the purpose of this chapter. Rather, this chapter will investigate how we can regress over similar incidents or tasks. We take "similar" to mean various forms of bipedal locomotion such as: walking, jogging, or running.

★★★

The study of powered prostheses sits within the larger emerging domain of rehabilitation robotics, where automation assistive machines (AAM), such as powered wheelchairs, neural prosthetics, and exoskeletons, play an ever increasing role in reestablishing locomotion in people who have lost it due to disease, accident, or war (where, e.g., neural prostheses are widely used for veterans of armed conflict) [1–4]. Beyond the replacement of limbs there are also many neurological and orthopedic disorders (e.g., multiple-sclerosis, stroke, Guillain-Barre syndrome, and cerebral palsy), which also reduce or eliminate voluntary recruitment of muscles. Such loss diminishes or renders impossible the performance of motor tasks or maintenance of muscles, connective tissue, and metabolic systems that depend on muscle activity for their function and integrity.

We seek to ease voluntary muscle recruitment by also sharing control of AAMs with the motor-impaired individual and an auxiliary system—see

Powered Prostheses
http://dx.doi.org/10.1016/B978-0-12-817450-0.00004-3

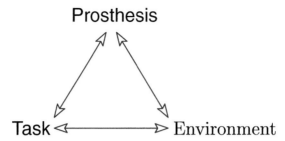

Figure 4.1 The interaction triangle between the device (prosthesis), its operating environment and the task it has been set to perform. Together they form the operating interaction triangle.

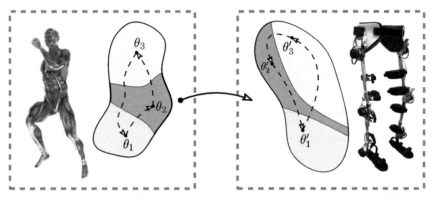

Figure 4.2 A simplified illustration of a human locomotion control manifold on the left, and its correspondent manifold, for the AAM on the right. A control path $\theta_{1\to3}=\theta_1 \to \theta_2 \to \theta_3$ is shown which corresponds to some sequence of activities (e.g., running → walking → jogging) where each parameter set θ lives on some sector of the activity-manifold (indicated by its color). The task at hands seeks to transfer the same control behavior to the AAM, so that $\theta_1' \to \theta_2' \to \theta_3'$ as closely as possible gives rise to the same kinematic behavior as $\theta_{1\to3}$.

Fig. 4.2 for an illustration of this. Where we augment current control approaches with that of a learned model (see Fig. 4.3 for deeper understanding about what sort of data we are considering) acting in parallel or in some cases; instead of the individual (i.e., a trade-off between complete AAM control autonomy or control of the AAM shared with the user), which would relieve her of significant control load, while still being in overall command of her locomotion. Within the clinical context of AAMs, machine learning (and in particular Bayesian nonparametrics) currently plays a limited, scarcely visible role [1]. Though AAMs are crucial in facilitating the independence of those with severe motor impairments, patients

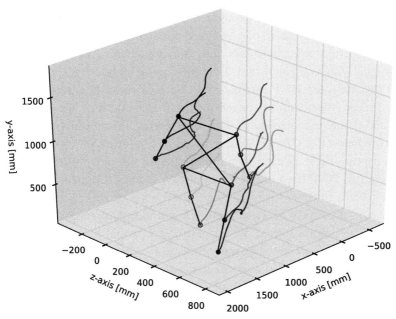

Figure 4.3 Motion capture (MOCAP) observation can be used to learn the temporal evolution of human dynamics. Shown is a subject performing a walking motion, in which the exogenous marker coordinates have been plotted over time as the activity evolves. *(Generated from observations used in [72]).*

exist for whom the control of these devices remains an insurmountable hurdle.

We have drawn upon one domain of mathematics and statistics in particular, in respect to tools which we perceive will become useful (computational complexity allowing) in future AAM control systems: Bayesian nonparametrics (BNP). The quantitative reasons for this will become clear as the chapter progresses, but we offer a brief discussion as to the qualitative notion of using BNP as part of the AAM control ecosystem. Nonparametric models constitute an approach to model selection and adaptation, where the size of the models are allowed to grow with data size. Comparatively, parametric models only use a fixed number of parameters—but what if our parameterization is not enough?

Any model selection will always be subject to model structure constraints, which is to say that even if we found the most optimal model $M \star$ with an inappropriate structure, it will *never* be able to accurately model the phenomena under study. Concretely, consider an example, that we want to model some second-order Markov process, but we are careless in our model structure selection and only specify a model with first-order Markovian dynamics. Then, irrespective of how good our inference is, we *cannot* capture those second-order dynamics with our models. And neither could a BNP model. But compared to a parametric model, under an appropriate model structure, the BNP model would be able to adapt the parameter space as more information is passed to the model, and thus appropriately grow (and shrink if so necessary). This a parametric model cannot do, so if some complex set of samples were received in the future, it would not be able to capture those dynamics faithfully. Therein lies our justification for supposing that BNP has a role to play in AAM control design.

Any patient fortunate enough to be fitted with an active prosthesis or orthosis, will effectively be a giant sensor, continuously passing new information to the control system. We suggest that it would be a poor design indeed, that does not take advantage of this rich spatiotemporal and high-dimensional information, in a way that allows the AAM not just to adapt, but to adapt to the individual user, and so offer truly personalized rehabilitation. This chapter therefore explores the interface between machine learning, biomedical engineering, and rehabilitation robotics, with the hope of enabling prostheses the freedom to automatically move and *feel* like a real leg for the user.

Specifically, we focus on the area of prosthesis velocity transition, which remains an open question, that is, how to make the prosthesis accelerate and decelerate like a healthy limb? Current approaches rely predominantly on optimal control methods for efficient regulation and adaptation. Few approaches rely of learning from human observation, let alone incorporating this into a control schema. Herein we present, *locomotion envelopes* which combined with impedance control, provides a powerful method for adaptive control, as well as a mechanism for smoothly transitioning between self-selected velocities.

1 Related work

Lots of people could benefit from a powered ankle. To this end, active prostheses[a] produce positive work during walking and inject energy to the gait during push off and propulsion. Powered ankle-foot prostheses can accommodate fast walking beyond the capability of their passive counterparts, and recent innovations and advancements have led to the development of

---

[a] These results are applicable to orthoses as well, but that is outside the scope of this chapter.

various devices for physical assistance and human locomotion restoration [5]. The design of a prosthesis is two-fold; one part is mechanical, the second is the control of the mechanical. In this review we focus on the latter.

Curiosity about control schemas date back several decades. Indeed, early work on powered transfemoral (above-knee) prostheses was undertaken by [6–8]. But, as noted earlier, control of these devices have two sources of origin (broadly speaking); those that explicitly seek to control prostheses for human ambulation, and those that seek a biomimetic solution for bipedal robots. Virtues and vices can be found in both domains, and we shall discuss both in turn, starting with the former. For emphasis, recall that we are *only* interested in strategies that relate to velocity adaptation and regulation, *not* the general field of prosthetics control strategies as reviewed by [9] and [5]. Indeed, as noted by [10] "available controllers for powered transfemoral prostheses cannot generalise across different walking speeds."

The most common solution used for control of active prostheses is to match the torque-angle profile of a healthy human ankle at the powered prosthetic joint. [11–14] describe a finite-state impedance control approach to control a prosthesis during walking and standing. This control method often leads to a few fixed walking speeds but lacks the adaptability required to seamlessly change the velocity or step size following the user's intention. In addition, gathering human experimental data at different speeds is limited to a few discrete values and it is difficult to perform numerous experiments to derive walking data at intermediate walking speeds or step sizes [15–17].

In order to address the adaptation problem of this control method [18] proposed a reflexive neuromuscular model with positive force feedback. They showed that the proposed method can adapt to changes in the walking speed and floor inclination. Further, Markowitz et al. [19] proposed a neuromuscular reflexive model with speed adaptation for a powered ankle-foot prosthesis and tested it at three walking speeds (0.75 m/s, 1.0 m/s, and 1.25 m/s). Hargrove et al. [20] extracted electromyography (EMG) signals from a patient, who had undergone targeted muscle reinnervation surgery, and then used these to provide a robust control mechanism for an ankle and knee prosthesis to walk with a fixed speed on level ground, stairs, and ramps with a 10 degree slope. Lenzi et al. [21] demonstrate a control approach for a transfemoral prosthesis, which regulates the ankle and knee joint torque by estimating the walking phase and speed. A two-dimensional lookup table with a low-pass filter was used to encode torque-angle curves for two-walking speeds. Quintero et al. [22] use a proportional-derivative controller with virtual constraints, which used a human-inspired phase variable to adapt to speed variations. The authors themselves note that their method could not quite compensate for strong nonlinearities in the ankle dorsiflexion and consequently produced control errors when this happened.

Now, Ferreira et al. [9] suggest that active powered prosthesis control falls into four categories including echo control, finite state impedance control, EMG-based control and central pattern generator (CPG)-based control. Conversely, Lenzi et al. [10] suggest that effective speed adaptation has been successful using two approaches. The first one, by Herr et al. [23], proposes a method that mimics human muscle reflexes. This allows the prosthesis to adopt velocity adaptation by virtue of changing the torque output without actually "measuring the walking or cadence" [10]. Though demonstrating very impressive results, there are some drawbacks. Specifically, because their guiding metric is the metabolic cost of transport, for five different velocities, across which they regress, it is difficult to ascertain how well their control schema transitions between velocities. Secondly, using the same taxonomy as Lenzi et al. [10], concerns the usage of preprogrammed (alternatively pre-specified) ankle-torque profiles. In order to cope with velocity variations, [24] modulated the ankle trajectory in time and amplitude, allowing their subject to transition between velocities. But this is a parametric method, relying on look-up tables for fitting without uncertainty bounds. Finally, Lenzi et al. [10] propose their own method which imitates the basic velocity adaptation mechanism used by healthy (i.e., intact) legs. They employ quasi stiffness profiles (we also show diagrams of these in our result section) of an intact leg, which they directly encode into their controller, and then interpolate between them based on their intention estimation. Theirs is the work which is most similar to that presented in this chapter. But while they use a PD controller, we use an impedance controller, and probabilistic interpolation and extrapolation, which also gives us uncertainty bounds on our predictions. Further reviews of control strategies for lower extremity prostheses can be found in [25].

As with our method, Gaussian processes (GP) have been used before for similar purposes; Hong et al. [26] use the GP dynamics model (GPDM) [27]. The GPDM is a dimensionality reduction method which comprises a low-dimensional latent space with associated dynamics, and a map from the latent space to the observation space, it is an elegant model of dynamics that accounts for uncertainty in the model. Hong et al. [26] use it to create a low-dimensional representation of walking motion, extracted from 50 subjects. They do this for three different speeds. This is not a control scheme, but a rehabilitation method, which can generalize between the training speeds (however their reconstruction errors are high > 10%) and is used for gait training of subjects with hemiplegia. Lizotte et al. [28] instead make use of standard GP regression (GPR) to optimize gait for quadruped and biped robots. Though they do not deal specifically with velocity adaptation and regulation (rather environmental adaptation) their ideas are relevant to our discussion, as environmental adaptation is the next logical step for the method presented within. Similarly Yun et al. [29] used GPR to generate a model for gait pattern prediction for one speed (3 km/h). Though different to what is presented, it does suggest a validation of the method herein, since the paradigm remains the same (and they demonstrate an impressive array of results); theirs is a prediction in space; ours in time.

A basic requirement of any human locomotion controller, is the ability to change speed and gait. We have shown some example studies where speed variation control is studied, for active prostheses. For completeness we also review recent advances within the field of humanoid bipedal robotics, whose aim and scope is similar to ours, and where, potentially, our methods are also applicable. These methods too have the potential to produce new controllers for bipedal robots and to provide simulation platforms for lower-limb prostheses [30].

Central pattern generators (CPG) have been successful in human locomotion controllers [31], where they are responsible for producing the basic muscle activation rhythms and local reflexes that modulate the muscle activations. CPGs can also be used to adapt to the environment. This locomotion paradigm has been successfully applied to prosthetics, see [32,33] (however, neither is capable of modulating speed). CPGs and reflex-only models [34] have in common that they generate locomotion, and switch locomotion regimes and behaviors by transitioning between different sets of control parameters [30]. But as Song and Geyer [30] note, it is unconvincing "that humans store look-up tables of hundreds or thousands of low-level control parameters for all different environments and behaviours." Some CPG-based studies have overcome this by using high-level policies that modulate low-level control parameters. For example, Van der Noot et al. [35] demonstrate a energy-efficient neuromuscular model, which mimics human walking. They do this by combining reflexes and a CPG able to generate gaits across a large range of speeds. They demonstrate their approach on a simulation model of the 95 cm tall COmpliant HuMANoid platform (COMAN) robot. Their results show that they were able to simulate energy-efficient gaits ranging from 0.4 to 0.9 m/s.

One drawback of their method is that it requires optimization of the open parameters of their model, and what is more: for a fixed-time simulation (they use 60 s). Further, the most common control method used in humanoid robots is based on the notion of zero moment point (ZMP). In ZMP walking, the feet are kept flat on the ground while the knees are bent—this is not suitable for human walking [36]. Further, related to this discussion, is dynamic walking:

> *A theoretical approach to legged locomotion which emphasizes the use of simple dynamical models and focuses on behavior over the course of many steps, rather than within a single step, typically in an attempt to understand or promote stability and energy economy.*

As Collins [37] goes on to explain: dynamic walking builds on the *passive* dynamic approach by adding simple forms of actuation and control ([37], Section 1.2.2). Here, when we refer to *passive*, we refer to a form of locomotion which is designed such that the natural oscillation of the system (i.e., the robot) results in a gait—where the analogy is drawn to the pendulum wherein the natural oscillation of that system is expressed through the swinging of its pendulum. Fully dynamic walking (i.e., not passive) adds actuation and control to the system without overwhelming the natural dynamics of the design. In the pendulum analogy this could be, for example, adding a small amount of torque at the apex of the pendulum, to increase the angular velocity, and then allowing gravity to pull it down again, once it has reached its peak (and then continuing to add small amounts of energy to the system to maintain this new state). Dynamic walking remains an elegant approach to biped control and research. Consider for example the recent work by Gritli et al. [38] where the authors aim to control the chaotic dynamics, exhibited in the semi-passive dynamic walking of a torso-driven biped robot as it goes down an inclined surface. Though elegant and important in applications were energy efficiency is paramount, it does require a gravitational differential to move down an inclined surface—for *passive* dynamic systems. However, when such requirements are relaxed, some studies explicitly use the knowledge obtained in passive dynamic walking, for control of actuated dynamic walking [39], with the overall aim of promoting energy efficient control.

★★★

In this chapter, we introduce an adaptive walking control method for a simulated powered ankle-foot prosthesis (Fig. 4.4) using universal function approximation methods, combined with impedance control. The adaptive property helps the user to change the walking speed, and the

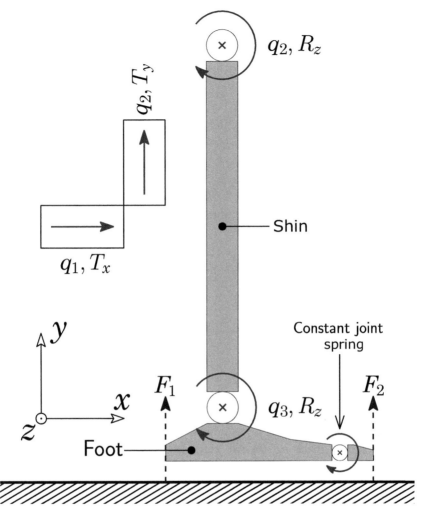

Figure 4.4 Diagram of prosthesis used in this chapter. In Fig. 4.4 Prosthesis-free body diagram—a schematic diagram is shown of the ankle-foot prosthesis with five degrees of freedom. In Fig. 4.4 T_x and T_y are translations along the x and the y axis. R_z is rotation about the z axis. F_1 and F_2 are ground forces. *(Adapted from [73]).*

nonparametric property enables the control scheme a higher level of precision as more observation are added. The proposed approach is general and can be applied to any ankle-foot prosthesis provided certain sensory information is available.

Initially we consider the sagittal degree of freedom (DoF). The shin is modeled with a rigid body connected to a planar floating base to simulate walking using recorded human walking data. The ankle-foot prosthesis and its interactions with the environment are modeled in the Robotran simulator [40] as shown in Section 5.

2 Locomotion envelopes

We present a novel control strategy for powered ankle-foot prostheses, using a data-driven approach, which employs a combination of Gaussian processes (GP) regression and impedance control. We learn the nonlinear functions, which dictate how locomotion variables temporally evolve using the aforementioned nonparametric method, and regress that surface over several speeds to create a manifold, per variable. The joint set of manifolds, as well as the temporal evolution of the gait-cycle duration, is what we term a *locomotion envelope*.

Current powered prostheses generalize poorly across speeds. Others that do have the capacity to exhibit several speeds, typically divide the gait cycle into several sequential periods. Each has independent controllers, resulting in many control parameters and switching rules that must be tuned for a specific walking speed and subject. It is not convincing that control strategies should rely on stored look-up tables of hundreds or thousands of low-level control parameters for different speeds. It is also unlikely that humans rely upon this approach, where a hierarchical control structure is more probable [41,42], such that gross motor-control precedes any low-level fine adjustments used for balance, for example.

★★★

The term *locomotion envelope*[b] refers to the joint set of multivariate regression surfaces which, appropriately applied, confers upon the user the ability to synthesize natural and robust bipedal locomotion (that same envelope also contains learned temporal regression functions, controlling the evolution of stride duration, across a sought range of gait speeds). We seek to regress physical properties such as joint angles, ground reaction force (GRF), and moments (GRM), from very sparse observational data. This can be viewed as a learning problem, where we are interested in learning the multivariate regression manifolds of the aforementioned properties.

There are a multitude of options, which one could use to construct these envelopes, we posit however that the most suitable is Gaussian process

---

[b] We derive the term from aerodynamics in which the flight envelope, service envelope or performance envelope of an aircraft refers to the capabilities of a design in terms of airspeed and load factor or altitude. The author will submit an aviation bias here: he is an aeronautical engineering by training.

regression. There are a number of reasons for this, some of which we expand upon here:

- First, given observations and a kernel, the posterior predictive distribution can be found exactly in closed form [43].
- Second, by nature of its construction, expressivity is considerable, allowing us to incorporate a host of modeling assumptions and domain knowledge.
- Finally, as noted by Rasmussen and Williams [43]; given a fixed kernel, the GP posterior allows us to integrate exactly over competing models, hence overfitting is less of an issue than in other candidate methods.

A Gaussian process is a method for universal function approximation, that is, some realization of a GP, with some kernel, is arbitrarily close to the function under study, to within some norm. Thus we can approach the multivariate function learning problem by placing a prior distribution on the regression function using a GP [43]. With a GP we can define a distribution over functions

$$f(x) \sim \mathcal{GP}\big(m(x), k(x, x')\big) \qquad (4.1)$$

parameterized in terms of a mean function $m(x)$ and a covariance function $k(x, x')$, with a multivariate observation denoted by $\mathbf{x}$. A GP is fully specified by these two functions. For a much deeper treatment of the GP, refer to the GP handbook by Rasmussen and Williams [43].

2.1 Noise modeling

One of the largest assumptions in GP modeling is that of the inherent noise in the model. *Why?* Because standard GPR assumes that input locations are noise-free [43]. The outputs however follow a homoscedastic (meaning "the same variance") noise process. Lets recall precisely the model using the standard linear model as a reference:

$$f(w) = x^T x \quad \text{s.t.} \quad y = f(x) + \varepsilon. \qquad (4.2)$$

Above, $\mathbf{w}$ are the weight parameters of the linear model and y is the observed target value and $\mathbf{x}$ our input vector. With GPs then, we assume a homoscedastic additive noise model, also seen in Table 4.1, specifically:

$$\varepsilon \sim \mathcal{N}\big(0, \sigma_n^2\big) \qquad (4.3)$$

in which σ_n^2 is the variance of our noise, and is assumed the same for all dimensions of x. The nature of this assumption cannot be understated, as

Table 4.1 Different noise models.

Model	Functional form
Additive noise	$y = f(x) + \varepsilon$
Linear noise	$y = a + b\varepsilon$
Posterior nonlinear noise	$y = g(f(x) + \varepsilon)$
Heteroscedastic noise	$y = f(x) + \varepsilon g(x)$
Functional noise	$y = f(x, \varepsilon)$

Rasmussen and Williams [43] explain: this noise assumption, together with the model, explicitly gives rise to the GP likelihood model—see Ref. [43] for details.

Under a different model, we would *not* receive this simple likelihood model. But, that is neither here nor there, rather, what is important is the other large assumption that we make with GP modeling: that x is uncorrupted. For many real problems this is not realistic. Even in our own experiments, where data was collected under rigorous test conditions, inputs are not noiseless. Indeed, there is no such thing as noiseless measurements, since we necessarily have finite precision with any available sensor. A noiseless variable is merely a theoretical construct. Going back to GPs, some interesting recent work which does consider noisy inputs is that by McHutchon and Rasmussen [44] where the authors present a simple GP model for training on input corrupted by i.i.d. Gaussian noise. Consequently they consider a model of the form:

$$y = f(x + \varepsilon). \tag{4.4}$$

This is unlike any of the noise models found in Table 4.1. In sum, the two limiting assumptions about GP regression noise are

- measured observations are noise-free;
- output points are corrupted by constant-variance Gaussian noise.

But as said, this is not a particularly realistic state of modeling. As Rasmussen and Williams [43] note, we do not have access to function values themselves, but only noisy versions thereof. The assumptions noted above work for many datasets, but for others, either or both of these points are invalid and can lead to poor modeling performance [44]. This discussion is relevant since we are proposing an anthropomorphic control system, which will interact with a human subject. Consequently it is important that noise model assumptions are discussed.

2.1.1 Complex noise models

In Table 4.1 we gave some examples of more complex noise models, models which do not follow the assumptions imposed by the GP model. We have discussed additive noise (assumed under the GP) and also homoscedastic noise. Lets go a step further and consider hetereoscedastic noise. The difference with this noise model is that it assumes that there is a direct dependence between the signal characteristics and the unwanted noise model [45].

Woodward et al. [45] explain that many instruments and processes (indeed most) have some component of hetereoscedastic noise such that their noise characteristics are dependent on signal characteristics. One can see immediately the consequences for biological control signals. First, it follows that an assumed hetereoscedastic noise model is highly personalized and not independent of the sensor modality. Second, imposing the assumption of this model would allow us to perform more accurate analysis, should we be able to estimate the noise signal with high accuracy, and make predictions with high accuracy (though we can no longer employ a closed-form posterior predictive distribution). Such a model is proposed by Goldberg et al. [46] where the dependent noise model is also modeled as a GP [$g(\cdot)$ in Table 4.1].

Suppose instead that the noise is *not* Gaussian, then what? Such a scenario was entertained by Snelson et al. [47]. Therein they explain that it is indeed somewhat simplistic to assume Gaussian noise. To overcome this, they "warp" the non–Gaussian noise input space into, for example, the log-space, and then the assumption is imposed that *this* space has Gaussian noise and can then be well modeled by a GP, with its simple construction. Though this may seem like a fanciful way to get around, or approximate, a rather difficult problem, Snelson et al. [47] explain that this is common practice in the statistics literature, so it is not without foundation. In a similar vein is the paper by Rasmussen and Ghahramani [48] where they discuss different noise variances, in different parts of the input space.

Consider the problem: suppose an input space is given, but it is large (high-dimensional) and may have discontinuities, a stationary covariance function will not adequately capture the latent function governing this space [48], and this naturally also includes the latent noise model. To tackle this, the authors present a mixture of experts model, but where the individual experts are GPR models (each with a different kernel). This allows the "effective covariance" [48] to vary with the inputs. This also means that the noise model can vary with the input, which is why this is an important contribution for domains which could have complex noise behavior.

2.2 Gait cycle stride-time regression

Having discussed noise models for the Gaussian process model, we now move onto another area of investigation: stride-time. We define stride-time as the time it takes between a foot leaving the ground, and that same foot touching the ground again. Naturally, this time incident will reduce with speed, as a consequence of bipedal locomotion. But as we go on to demonstrate, stride-time and speed, are not linearly correlated, and moreover demonstrate a high-degree of variability. As an example consider the curves in Fig. 4.5. Here we show all the extracted cycles of the plantarflexion angle, plotted against the time it took for the variable to complete one full cycle, before repeating itself again.

Much in the same way that we extract single-cycles from the long time-histories of observations for space variables (like the one plotted in Fig. 4.5), we do the same for the time-variable (see Fig. 4.6).

Because we are extracting a short quantity (stride-time) from a large one (the full time-history of a trial) we are left dealing with a lot of data. One of the primary assumptions of this work, is the notion of "atomic" locomotion envelopes, that is, it is enough to do inference over the primary building block (one gait cycle) of locomotion, rather than the whole time-history. Hence, we adopt the same approach here: gait cycle duration is measured for each trial, then an average is taken for that trial and hence speed, and

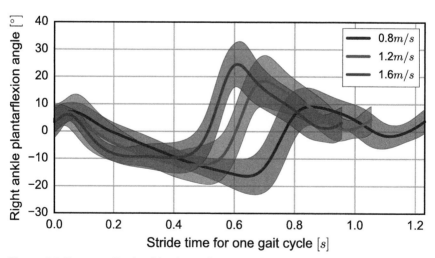

Figure 4.5 Un-normalized ankle plantarflexion angle with ± two standard deviation during normal walking at three different speeds. Depiction demonstrates on the horizontal axis the stride-time required to reach various parts of the gait cycle. It is clear that just after the middle there is large acceleration in angle.

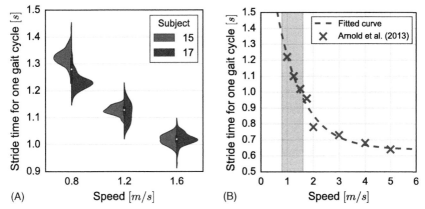

Figure 4.6 (A) Depicting comparative stride duration violin plot. (B) Curve fitted to [16] stride duration observations. Left subplot depicting the comparative duration times for subject 15 and 17 from [16]. Right plot showing the logarithmic stride duration trend fitted to data from [16]. *(Adapted from [73].* Note that these two subjects were chosen because they showed the greatest similarity in terms of physiology as well as weight (mass). While two completely different subjects could have been shown, interest lies in finding common denominators across subjects).

then scaled according to body mass and height. For further details on this process see the primary studies from whence our data was taken [15–17].

Continuing, each envelope needs to scale each manifold by time as the prosthesis accelerates or decelerates. This is a nonlinear relationship, which, like the manifolds, also needs to be approximated. This, however, is a simpler task than the preceding one. To aid our exposition and comparison, we employ the study by Arnold et al. [16], wherein motion capture data was collected for five subjects walking and running on a force–plate instrumented treadmill (for a good idea of the setup, refer to the original study [17]). The subjects were all experienced long-distance runners who reported running at least 30 miles per week. The subjects walked at 1.00 m/s, 1.25 m/s, 1.50 m/s, and 1.75 m/s, and ran at 2.0 m/s, 3.0 m/s, 4.0 m/s, and 5.0 m/s.

The violin plot shown in Fig. 4.6A shows the distribution of extracted stride duration, across two discretized variables, speed and time, enabling the display of the comparative distributions. While it is clear that the mean of the duration kernels, for both subjects, disagree for speeds 1.2 m/s and 1.6 m/s, they diverge at 0.8 m/s. Why the divergence at 1.2 m/s and 1.6 m/s? It is likely to be more an artifact of the individual gait cycle—a physical manifestation of subject 15's overall slower gait at lower speeds. This is further reinforced if we consider their respective physiologies (see the full set of data by Moore et al. [17]), they show little divergence in that regard: both

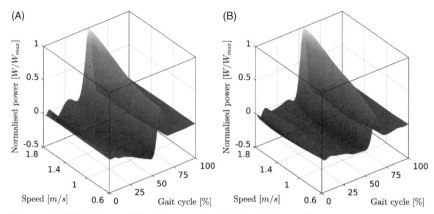

Figure 4.7 (A) Subject 6 (B) Subject 10 Manifolds depicting the monotonic increase in power usage, with increase in speed, over one gait cycle, for subjects from the Moore dataset [17]. Surfaces were found using GPR and shown is the posterior mean surface, where the uncertainty surfaces have not been included to reduce clutter and increase clarity. *(Adapted from [73])*

weigh approximately 85 kg, and are both circa 1.80 m tall—as well as of similar age. Hence, their anatomies suggest that they should manifest a gait pattern of high similarity, and they do, but not in the higher speed ranges for reasons discussed. The same figure also shows the speed region (in gray) investigated by Moore et al. [17]. Using a simple fitted function (inverse exponential) we are able to get a good estimate of the stride duration at our test points.

2.3 Analysis of human ambulation

Having established our approach for nonlinear regression, it is pertinent to be able to understand precisely what features we seek to regress. This calls for an understanding of human ambulation. Hence in this section, we verify common and general physical features observed in all healthy subjects as they walk faster, again drawing from the dataset produced by Moore et al. [17] (which also limits our verification to the number of test subjects). Consider Fig. 4.7 wherein we demonstrate that the power at ankle increases monotonically with an increase in speed. The yellow part of each subfigure shows when the normalized power is approaching its peak. The clear trend, for all subjects, is an increase in power injection, with an increase in speed. The injection happens at approximately the same place in the gait–cycle, but the difference is its magnitude, which as can be seen; increases with speed.

The power is computed by deriving the product of ankle moment and ankle angular velocity and then normalizing. In more detail, work is given by

$$W = F \times d, \tag{4.5}$$

where F is the applied force and d is the distance moved in the direction of the force. Power, taken as a function of time, is the rate at which work is done and is expressed by

$$P(t) = \frac{W}{t}, \tag{4.6}$$

In mechanical systems, such as ours, we consider the combination of forces and the movements they give rise to (such as the ankle displacement). One can think of the ankle's movement as a rotation about a point, we can express its power as

$$\boldsymbol{P}(t) = \tau^{\mathrm{T}}\boldsymbol{\omega} \tag{4.7}$$

where τ is the torque and ω the angular velocity, about the same point. Both quantities we can extract from the measurements, when inverse kinematics has been applied.

Fig. 4.7 confirms what we already know [49]: increase in walking speed occurs as a result of increase in the power exerted by humans at the ankle. Moreover, human observation suggests that this feature is unique to the ankle joint where a clear increase in power at push–off can be observed. This phenomenon is not present when looking at the knee or the hip joints in the sagittal plane during human walking. Because this is a universal feature that can be observed in human observations, it can furthermore be reproduced in powered ankle-foot prostheses. That being said, this has been investigated before. See for example the work by Farris and Sawicki [49] where the authors find that while "steady locomotion at a given speed requires no net mechanical work, moving faster does demand both more positive and negative mechanical work per stride." They also confirm that the ankle joint is chiefly responsible for the maximum percentage of total average positive power contributed, when undertaking locomotion, followed by the hip and then the knee joint. This is for all levels of human ambulation.

2.4 Experimental data

The experimental data used for this study, comes from the excellent set of experiments conducted by Moore et al. [17]. Therein the authors collected a rich gait dataset with the help of fifteen subjects, walking at three speeds (as shown in Fig. 4.5) on an instrumented treadmill (a more advanced form of treadmill, with independent right and left tracks, and universal perturbation functionalities). They explain that each trial consisted of 120 s of normal walking and 480 s of walking while being longitudinally perturbed during each stance phase with pseudo-random fluctuations in the speed of the treadmill belt. We are primarily interested in the normal walking observations, but note that the methods discussed herein would also form an interesting study if applied to the perturbed data as well. The details of the dataset are such that they contain: full body marker trajectories, ground reaction loads (labeled under the aforementioned gait events), two-dimensional (2D) joint angles (i.e., those from the sagittal plane), angular rates and joint torques. All of these were collected at 0.8 m/s, 1.2 m/s, and 1.8 m/s, for each subject—for further details see the original study [17]. An exposition of the raw ankle plantarflexion angle, over one gait-cycle, is presented in Fig. 4.7, where the joint-angles were found using inverse kinematics. For manifold learning, in succeeding sections, all trajectories are normalized to have the same discretized length in time—unlike what is shown in the example in Fig. 4.5. Further, we broadly follow the same protocol presented in section "Processed data" of Moore et al. ([17], p.16), with minor edits.

3 Simulation setup

Like many before use we employ impedance control. The justification for this is trivial; because our prosthesis is inherently interacting with its surrounding environment, we need to model that mechanical interaction. The scheme, which shows most promise in this regard, is impedance control [50]. For the sake of completeness we review the fundamental ideas further.

But before moving onto the minutiae of impedance control, it is worth considering the information flow considered thus far. We have demonstrated how observations are received, extracted and regressed using Gaussian process regression. This information flow is summarized in Fig. 4.8. As Fig. 4.8 shows, there are a number of steps that need to be undertaken to receive a locomotion envelope over multiple velocities, which can be subsequently used in an impedance controller as described further.

3.1 Impedance control

Impedance control (IC) is an extensive control schema, in which a mass-damper-spring relationship (see Fig. 4.10) between a position and force is established ([24], Section C) (Fig. 4.9). Much like when a robot interacts

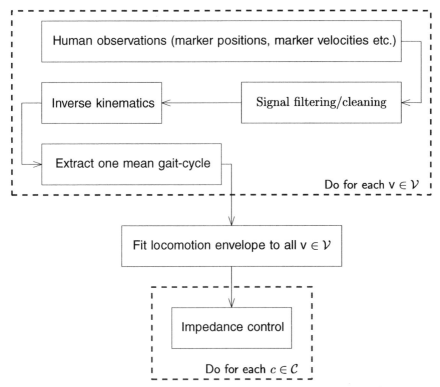

Figure 4.8 The information flow involved in using locomotion envelopes for control. The necessary training observations are extracted and preprocessed in the *blue* box. First, relevant signals are extracted such a MOCAP marker trajectory observations. Alternatives to MOCAP data could be, for example, accelerometry observations, taken from relevant positions of the lower body. Whichever observation set is used, these are subsequently cleaned using a variety of filtering and data cleaning processes, relevant details of which are found in ([17], p. 16). Thirdly, inverse kinematics is used to find the relevant endogenous angles such as the plantarflexion angles, since we cannot measure these directly. Inverse kinematics is also used to find important gait reference events such as toe-off and heel-strike timings. Having extracted gait events which segment the entire gait-cycle, we are in a position to enumerate all of them (i.e., $\cup_{i=1}^{N} c_i = C$) and subsequently find an average of the whole set of gait-cycles and standard deviation of one time-history (remember that we are only interested in one time-history of gait, in other words; one cycle). We repeat all of these steps, for all velocities v in the dataset (ideally v should be large to cover a large dynamic range of locomotion). Finally, having extracted mean gait-cycles for multiple velocities, we pass these to the locomotion envelope formalism, and subsequently the onboard impedance controller. Once on the impedance controller side (the *red* dashed box) the controller operates over an unbounded set $|C| = [0, \infty)$, dynamically adapting throughout usage.

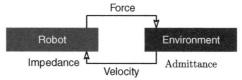

Figure 4.9 Understanding impedance control. Posit impedance as a dynamic operator that determines an output effort (force) given an input flow (velocity). Whereas an admittance is a dynamic operator that determines an output flow (velocity) given an input effort (force).

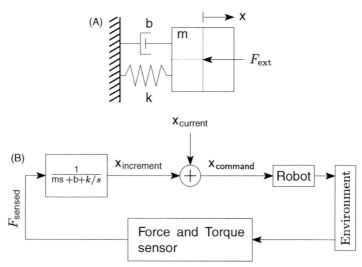

Figure 4.10 An example of a simple one DoF system (A) and its corresponding impedance controller (B). (A) A robot end-effector, for example, a lower limb actuator model, idealized as a pure mechanical system. Mass, damping, and stiffness parameters, are given respectively by m, b and k. An external force is applied, denoted by F_{ext}. (B) Control block diagram of the example impedance controller. Where the control command sent to the robot is $x_{command}$, $x_{current}$ is the current position and the required increment is $x_{increment}$.

with its environment, interaction forces result which, rather than being rejected, have to be accommodated as noted by Chan et al. [51]. The authors further note that to accomplish this, in addition to position control, force control is also required to accomplish the given task. It was explained elegantly by Mistry [52], where he says that in order to design controllers that can cope with environmental uncertainty we must treat the robot as an impedance and its operating environment as an admittance. This is shown and explained in Fig. 4.10, and relates back to our first treatment of this topic in Fig. 4.1.

First proposed by Hogan [53], IC yields a regime in which motion is commanded and controlled, and the "response for deviation from that motion" [51], resulting from the interaction force, is given in the form of an impedance (i.e., how much a robot resists motion when presented with an external force). This is desirable because it allows the robot the luxury of changing its effective dynamics in response to variations in its environment [24]. This can also be construed as the robot changing its resistance to its surroundings, which could be, for example, different types of ground surfaces or inclined and declined environments. Though attractive, one serious drawback is that in order for it to work properly, one needs incisive understanding of the force experienced by the robot in response to its environment. In simple impedance control, consider an example of IC applied to a simple model, which can be used to simulate the actuation motion of a lower limb. In this instance, we are really looking for a specific relationship between the externally applied force and robot motion. Impedance control generalizes the actuator so as to simulate a mechanical system, characterized by mass, damping, and stiffness. A simple instantiation of such a robot is given by the 1-DoF linear system in Fig. 4.10A. The equation of motion for this system is then given by

$$m\ddot{x} + b\dot{x} + kx = F_{\text{ext}} \tag{4.8}$$

The Laplace-transformed equation of motion is given by

$$\frac{V}{F_{\text{ext}}} = \frac{1}{ms + b + k/s} \tag{4.9}$$

where V is the Laplace transform of the velocity $\dot{x}$. For a block diagram of this control scheme see Fig. 4.10 - this example was inspired by Pham [54].

Having thus gained a firmer understanding of impedance control, consider now our application of the latter to our ankle prosthesis. Fig. 4.4 shows four revolute joints in total where three joints q_1, q_2, and q_3 are the floating base joints specifying the $x - y$ position and orientation of the shin in the plane (recall that for this study we are *only* operating in the sagittal plane). The fourth joint q_4, represents the planar motion of the ankle that is actively controlled during walking. The fifth joint q_5 is a passive toe with a fixed spring stiffness of 90 Nm/rad and damping of 3 Nms/rad that was added to better accommodate human foot kinematics. We let $q \overset{\Delta}{=} [q_1, q_2, q_3, q_4, q_5]$. With this in mind, we can state the equation of motion for this system

$$\boldsymbol{M}(\boldsymbol{q})\ddot{\boldsymbol{q}} + \boldsymbol{c}(\boldsymbol{q}, \dot{\boldsymbol{q}}) = \boldsymbol{\tau} + \boldsymbol{J}^{\mathsf{T}} \boldsymbol{F}_{ext} \tag{4.10}$$

where $\mathbf{M}$ is the mass inertia matrix, $\mathbf{c}$ is the vector of Coriolis, centripetal and gravity forces, $\mathbf{J}$ is the Jacobian matrix, and $\mathbf{F}_{ext}$ represents the ground reactions forces. The kinematics and dynamic parameters were extracted from each subject's MOCAP data as reported in [17].

The ground reaction forces and moment are applied at the foot's center of pressure, and the walking simulation consists of a swing phase where the ground reactions are zero and a stance phase where a part of the foot is in contact with the ground.

In order to show the functionality of the proposed nonparametric regression methods, in changing the walking velocity, inverse dynamics is performed on the inferred manifolds to illustrate acceleration and deceleration during transitions from one speed to another. These transitions can occur at any moment during the gait cycle. A sensor is attached to the center of the foot in the simulation to obtain important gait features such as step-size, step-frequency, velocity, and acceleration. Further, consider that the stance phase of walking can be divided into three distinct sub-phases. The first is controlled plantarflexion (CP) that occurs right after heel strike. This phase is followed by controlled dorsiflextion (CD) that occurs when the angle between the shin and ankle starts to decrease. Finally, this is followed by powered plantarflexion or the push-off phase that is the main focus of this chapter and it is where energy and power is injected into the walking gait. The event timings of swing phase starts and ends are measured in the simulation.

4 Prosthesis impedance controller

The prosthesis used in this chapter is the device developed by Ficanha et al. [55]. It is a prosthesis designed to "meet the mechanical characteristics of the human ankle including power, range of motion and weight" [55]. To allow for optimal placement of motors and gearboxes, transfer of power from the motors and gearboxes to the ankle-foot mechanism, we use a Bowden cable system. To control the prosthesis, impedance controllers in both sagittal and frontal planes were developed. We only consider the sagittal plane DoF. The impedance controllers used torque feedback from strain gages installed on the foot.

Two impedance controllers are required to control each motor independently. Each impedance controller uses an external position controller with an internal torque controller. The external position controller tracks a reference trajectory (generated through GPR) and uses the

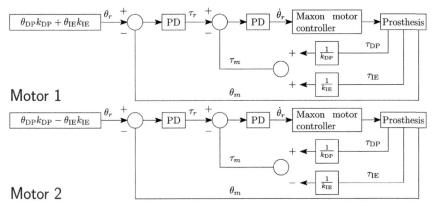

Figure 4.11 Impedance controllers for prosthesis motors. *(Adapted from [73]).*

angle feedback from the encoders on each motor θ_m. The block diagram of the impedance controller is shown in Fig. 4.11. The motors actuate the ankle in dorsiflexion-plantarflexion (DP) and inversion-eversion (IE) directions using Bowden cables that form a differential drive mechanism.

The output of the external position controller is the desired torque to be generated by the motor τ_r and is the input to the internal torque controller. The internal torque controller uses torque feedback from strain gauges mounted on the foot. The strain gauges provide both the torque in dorsiflexion-plantarflexion (τ_{DP}) and the torque in inversion-eversion (τ_{IE}). Torques τ_{DP} and τ_{IE} are calculated from the strain gauges voltage outputs as described in previous work [55]. Due to the differential drive nature of the mechanical setup, the sum of τ_{DP} and τ_{IE} is used for the reference torque (τ_r) for one of the motor controllers, and the difference of τ_{DP} and τ_{IE} is used in the other motor controller. Similarly, the reference trajectory θ_r for one of the motor controllers is the sum of the desired ankle angles in DP (θ_{DP}) and in IE (θ_{IE}), and for the other motor controller, the reference trajectory θ_r is the difference of the desired ankle angles in DP and in IE. The torque controller output is the desired motor velocity $\dot{\theta}_r$ and used as input to the Maxon motor controllers [55]. The impedance controllers were implemented with a real-time frequency of 200 Hz while the Maxon motor controllers used a proportional-integral controller running at 53 kHz. The plus-derivate (PD) block shown in Fig. 4.11 is used to control the prosthesis in both the frontal and sagittal planes. As noted, we only consider the sagittal plane. For further details see the work by Ficanha et al. [55].

5 Empirical evaluation

Our method inherently relies on the generation of trajectories which our controller can follow, to a high degree of accuracy. These trajectories contain the nominal phases; CP, CD, and push-off, to simulate as closely as possible inferred gait patterns at test velocities outside our training data. The trajectories are extracted from the inferred manifolds in the envelope, to show cyclic behavior at any speed and acceleration-deceleration transitions at any time during the gait cycle. These are the two key properties that are highly desirable to endow upon powered ankle-foot prostheses; speed adaptation and control repeatability. This is illustrated in Fig. 4.13, where trajectories extracted at different velocities are shown in different colors. During human experiments the subject may walk with a certain self-selected step size and speed that can be both extracted from this manifold.

In this section we apply our methodology to a set of experiments, demonstrating the utility in using locomotion envelopes for synthesizing robust locomotion. First, however, we provide a brief exposition on our choice of kernels. Their selection is paramount for accurate application of these methods. The reasons for this is intuitive. Like so many problems in machine learning, we are interesting in the underlying mechanism that gave rise to the observations that we are trying to model. In this chapter, we have chosen to employ Gaussian processes, as our weapon of choice. Furthermore, we have also noted that the Gaussian process is defined by $m(\cdot)$ and $k(\cdot,\cdot)$ and since we will be setting the mean function to zero, we are left with a process that only depends on the covariance function.

5.1 Kernel design

The use of kernel-based nonparametric GPs has been alluded to in previous sections. Good performance for these methods is highly conditional on the choice of kernel structure *as it encodes our assumptions about the function which we wish to learn* ([43], Section 4). Typically this choice can be somewhat difficult, and methods have been proposed for automating the selection process [56], in our case, however, we have substantial prior information regarding the nature of our multivariate regression surfaces.

The kernel function $k(x,x')$ determines how correlated or similar our outputs y and y' are expected to be at inputs $\mathbf{x}$ and x'. These inputs could be, for example, time-dependent joint angles at a set velocity at some part

Table 4.2 Compositional kernels used for kriging.

Target	ARD	Periodic	SE	RQ	M_{52}
Joint angles	✓				✓
GRF (x, y)	✓	✓	✓		
Marker Positions (x, y)	✓			✓	
GRM (x, y)	✓	✓	✓		

of the gait-cycle, respectively. By defining the measure of similarity between inputs, the kernel determines the pattern of inductive generalization.

In Table 4.2 we have summarized the kernel structure for the present variables in the envelope. These are, in no particular order: GRF (x, y), GRM (x, y), marker position (x, y) as well as ankle and knee joint-angles. In Fig. 4.12 we compare some different kernels on the same regression problem, for a comparison of outcome.

Fig. 4.12 shows the posterior mean function without uncertainty bounds. To begin our discussion, most interesting is Fig. 4.12B because although it generalizes poorly in the velocity direction (recall: we are trying to infer locomotion parameters across a range of velocities) it overfits the training data. The squared exponential kernel with isotropic distance measure is given by

$$k_{\mathrm{SE}}(x, x') = \exp\left(-\frac{\|x - x'\|^2}{2l^2} \right) \qquad (4.11)$$

where it can be seen that the length scales do not scale the inputs according to relevance of that input dimension. Unlike the other kernels we investigate, which all use this type of dimensionality scaling, Fig. 4.12 does not and it can be seen that the temporal dimension dominates the others (i.e., the evolution of the gait cycle).

As we are only considering one period of the gait-cycle, as this is one of our primary modeling assumptions, we use periodic kernels where beneficial to do so, yielding functions of the form $f(x) = f(x + P)$ where the P is the period of the gait-cycle. At times the addition of a periodic kernel has no benefit to the predictive performance, in which cases it has been omitted. Further, manifestly it is clear that the function values of $f(\cdot)$ change faster, and more slowly, depending on which input dimension $x \in \mathbb{R}^2$ we are considering. Intuitively this means that directional changes are of no importance. This is too strong an assumption in our case hence why isotropic kernels are unsuitable for our application domain. Instead we employ automatic relevance determination (ARD), which appropriately scales the

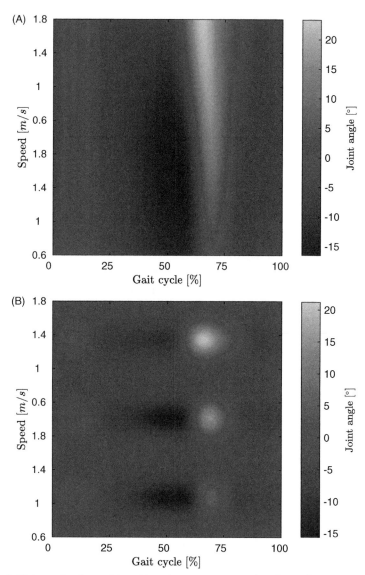

Figure 4.12 Investigating subject 10 and regression over the plantarflexion angle, the above plots depict different posterior predictive mean surfaces, using various kernel choices. As can be seen, aside an isotropic squared exponential kernel in part (B), there is little to differentiate the posterior mean surfaces, and they evaluate to approximately the same mean posterior. (A) M_{52} – ARD kernel. (B) Isotropic SE kernel. (C) SE-ARD kernel. (D) Compound kernel.

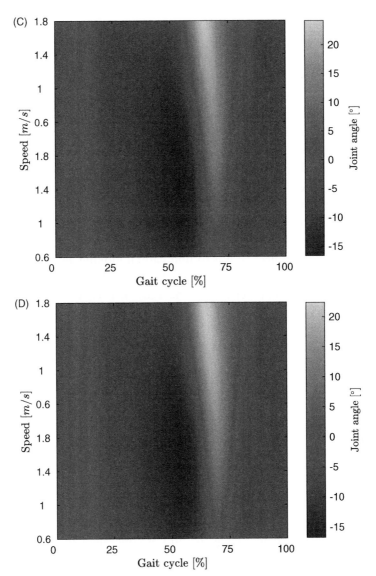

Figure 4.12 (*Cont.*)

inputs, thus determining the "relevance" of each dimension. Consider the ARD SE kernel

$$k_{\mathrm{SE}}\left(\boldsymbol{x}, \boldsymbol{x}'\right) = \prod_{d=1}^{D} \exp\left(-\frac{||\boldsymbol{x}_d - \boldsymbol{x}'_d||^2}{2l_d^2}\right) \tag{4.12}$$

where l_d is the length scale as a function of input dimension d. Note that small length scale value means that function values can change quickly, large values characterize functions that change only slowly.

The final part of our kernel design is rather more crucial as it concerns the innate periodicity assumptions of our data. While it is clear that gait-cycles are periodic, they are *not exactly* periodic. This is further reinforced by our usage of the mean gait-cycle (for one period). Thus to allow for realistic variations over time, by design we make our kernels locally periodic, by multiplying by a local kernel. This allows us to model functions that are only locally periodic, the shape of the repeating part of the function can now change over time [57]. We experimented with different local kernels[c], each of which makes different smoothness assumptions about our data. The final form for each variable group, is shown in Table 4.2, where our design objective was to find natural looking, numerically consistent (with experimental data) simulations. Finally, as example exposition, a locally periodic version of the RBF kernel, with ARD, that is, $k_{\text{PER}\times\text{SE}}(x,x')$, is given by

$$\prod_{d=1}^{D}\sigma_d^2\exp\left(-\frac{2\sin^2\pi\left(\frac{||x_d - x'_d||}{p_d}\right)}{l_{1,d}^2}\right)\exp\left(-\frac{||x_d - x'_d||^2}{2l_{2,d}^2}\right) \quad (4.13)$$

where $l_{1,d}$ is the length scale as a function of input dimension d, for the k_{PER} kernel as is the periodicity p_d. Where $l_{2,d}$ is the length scale as a function of input dimension d, for the k_{SE} kernel. Having considered our kernel choices, we are in a position to use them for experimental evaluation. Before that, consider again the brief experimental comparison undertaken in Fig. 4.12.

The numerical comparison of these kernels follow in Table 4.3 where the negative log-likelihood (NLL) has been computed.

Table 4.2 shows that as far as the NLL is concerned, there is not much difference between the kernels. Broadly they settle on the same posterior mean function, practically they do not, as the example with isotropic SE kernel shows. Without scaling the input dimensions according to their relevance, overfitting such as this will occur, hence the importance of using ARD for many problems. On the other hand, from a smoothness point of view, the posterior mean functions look broadly the same, save

---

[c] For details on the Matérn kernels with $v = 5/2$ (M_{52}), and the rational quadratic (RQ) kernel, see (§4.2).

Table 4.3 Minimization of the negative log marginal likelihood $\mathcal{L}(\theta)$ with respect to the hyperparameters and noise level, under the chosen kernel. Lower is better.

Kernel	NLL
Squared exponential with ARD	-3.1643×10^3
Isotropic squared exponential	-3.1139×10^3
Matérn 5/2 with ARD	-2.8275×10^3
Compound (Per-ARD with isotropic SE, multiplied w. SE-ARD)	-3.1785×10^3

for small differences, meaning that they are all viable candidates for our experiments.

5.2 Accelerating and decelerating

As stated at the beginning of this chapter, there is a gap in the literature concerning transitions between different velocities for powered prostheses, that is, few are able to smoothly transition between user-selected velocities. As this is a key capability that we wish to incorporate into future devices, our first simulation experiments seeks to emulate this behavior. Fig. 4.13 describes the scenario we are interested in; the different colored

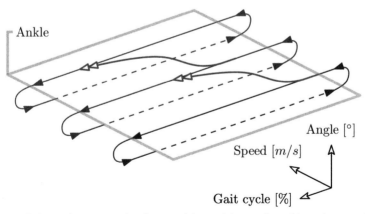

Figure 4.13 A mock-up scenario of potential transition paths, taking place on the inferred ankle manifold. Shown are possible gait-transition functions (double-headed cyan arrow paths). Trajectories of different color, indicate trajectories at different velocities. The last curve, from the coordinate frame seen, operates under the highest velocity. The vertical direction of the coordinate system is a reference to any of the posterior mean function plots, see example, Fig. 4.14, where the direction out of the page represents the target variable, and in this instance the target for the ankle is the plantarflexion angle.

Table 4.4 Velocities used for training and testing in the locomotion envelope formalism. Note that testing takes place well outside the range of the training data.

Data	Low [m/s]	Mid [m/s]	High [m/s]
Train	0.8	1.2	1.6
Test	0.6	1.2	1.8

paths represent gait-cycle at different velocities. A transition event seeks to smoothly go from one cycle to another, "smoothly" here is a misnomer, as what we mean to say is a transition that is realistic and anthropomimetic.

In this experiment, a representative path on the inferred manifold (the posterior mean function) is chosen to simulate three speeds and their smooth transitions. The inferred joint angles and ground reaction forces and moments are used in inverse dynamics calculations to illustrate walking with speed adaptation at speeds 0.6 m/s, 1.2 m/s, and 1.8 m/s. Hence, we seek to demonstrate our methodology simply by instantiating a simple forward simulation under our model and controller, to show that it is able to smoothly transition between gait-cycles as shown in the mock-up in Fig. 4.13. The model is trained on plantarflexion curves at 0.8 m/s, 1.2 m/s, and 1.6 m/s, the simulation transitions between simulated curves (through GPR) at 0.6 m/s, 1.2 m/s, and 1.8 m/s. This is summarized in Table 4.4.

The walking speed is measured by placing a sensor on the foot in simulation that illustrates the instantaneous position, velocity, and acceleration. Fig. 4.16 demonstrates accelerations using our methodology. Each speed profile is repeated over three cycles before acceleration to the next speed profile. It is interesting to note that the lowest (0.6 m/s) and the highest speed (1.8 m/s) were not available from the experimental data and they were inferred by GPR. Granted, many interpolation and extrapolation methods (e.g., support vector regression, neural networks, splines, piecewise linear interpolant, pure radial basis functions, etc.) are able to extrapolate well outside their optimization domain. But as has been noted, GPR, this of spatial regression, has particular benefits the primary of which is the uncertainty incorporation. Though not shown in this example, we do have uncertainty bounds on all our posterior mean functions in Fig. 4.14—see Fig. 4.15 for an example using the knee-flexion angle (we use this as an example, because it is not an active degree of freedom in our simulation, hence serves merely as a demonstration of spatial regression with GPs).

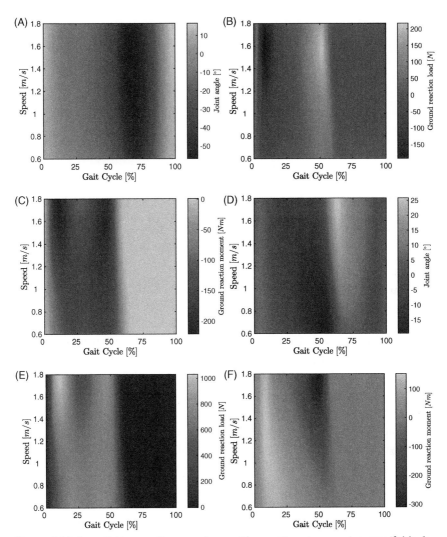

Figure 4.14 A partial locomotion envelope, with constituent regression manifolds, for subject 6 from the Moore dataset [17]. This envelope is used to instantiate a forward simulation. As before, for each manifold, training data was used at speeds 0.8 m/s, 1.2 m/s, and 1.6 m/s and GPR used to regress each variable in the envelope, to a speed domain of 0.6–1.8 m/s, over one gait cycle. (A) Knee flexion angle. (B) Ground reaction load in x. (C) Ground reaction moment in x. (D) Ankle plantarflexion angle. (E) Ground reaction load in y. (F) Ground reaction moment in y. *(Refer to Fig. 4.4 for the reference frame. For a summary of the velocities used for training and tests, refer to Table 4.4).*

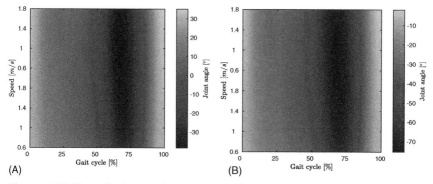

Figure 4.15 Knee flexion angle uncertainty bounds. Depicted are the posterior mean surfaces plus and minus the posterior uncertainty. Looking at the color bar, it can be seen that the variability in the uncertainty is considerable. (A) Posterior mean surface plus uncertainty surface. (B) Posterior mean surface minus the uncertainty surface.

Having extracted a locomotion manifold (see a partial envelope in Fig. 4.14 for a set of training points that lay appreciably far away from the training data (circa ± 15% for speed), we are in a position to synthesize locomotion sequences, consisting of several speeds, durations, accelerations and decelerations[d]. As can be seen from Fig. 4.17 and Fig. 4.18 we are able to demonstrate multiple properties of our method in this simulation. At the start of this chapter we emphasized the need for a prosthesis to be able to accelerate and decelerate, at the will of the user. In Fig. 4.16 we see that acceleration is taking place is a smooth fashion (for more robust evidence see accompanying video in the supplementary materials). Taken from the regression manifolds, we enable the controller to generate motion that is consistent with those manifolds, as well as the transitions between the cycles on the manifold.

Second, the gait-cycle path was tracked using concatenated gait-cycles (see Fig. 4.17), found from the learned (offline) multivariate regression functions. The gait is realistic and mimics subject six's gait well. More importantly, we demonstrate a physically sound gait, not just for GRFs and GRMs, but also joint-angle torques. The inferred joint trajectories and ground reaction locomotion envelopes are used to simulate speed transitions from slow to normal to fast walking speeds. The snapshots of the inverse dynamics simulations of different speeds are shown in Fig. 4.18.

---

[d] See the supplementary material at https://youtu.be/iU7hNKLUX7c

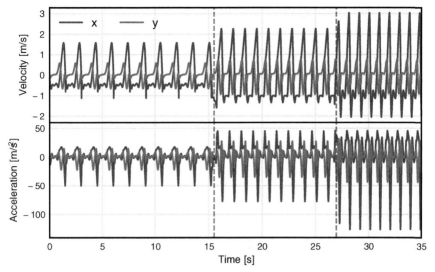

Figure 4.16 Forward simulation results for simulated transfemoral prosthesis, multiple walking cycles at each velocity, before transitioning to the next. Cartesian velocity and acceleration are shown of the foot, measured during two speed transitions, indicated by *(vertical dashed)* lines. *(Adapted from [73]).*

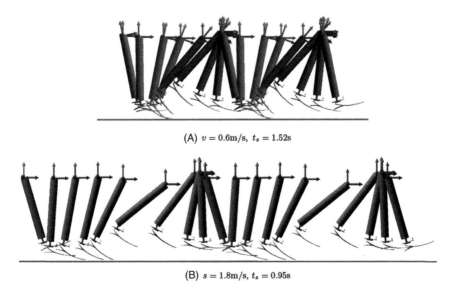

Figure 4.17 Long simulation experiments. Panels show gait cycle simulations for velocities outside the training data, where t_s is the stride time. These longer simulations, compared to Fig. 4.18, demonstrate that the method does generate the expected (i.e., realistic looking) gait and cadence. Again, see the supplementary material for a video depicting this gait sequence. (A) $v = 0.6$ m/s, $t_s = 1.52$ s. (B) $s = 1.8$ m/s, $t_s = 0.95$ s. *(Adapted from [73]).*

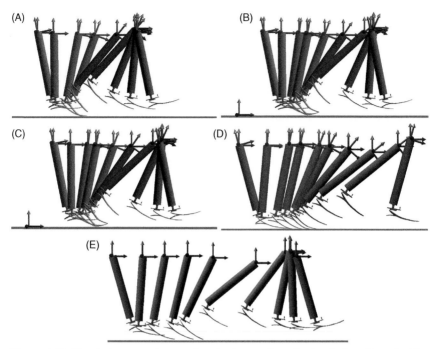

Figure 4.18 Simulation experiments. Panels show gait cycle simulations for velocities outside the training data, where t_s is the stride time. (A) $s = 0.6$m/s, $t_s = 1.52$s. (B) $s = 0.8$ m/s, $t_s = 1.23$s. (C) $s = 1.2$m/s, $t_s = 1.11$s. (D) $s = 1.6$m/s, $t_s = 1.1$s. (E) $s = 1.8$m/s, $t_s = 0.95$s. *(Adapted from [73]).*

5.3 Torque-angle relationship at test points

From simulation we can extract torque-angle curves at the knee and ankle angles in the sagittal DoF, and compare them to empirical results found in Moore et al. [17]. In Fig. 4.19 we have super-imposed curves for speeds $s = 0.6$ m/s, 1.8 m/s, to the experimental ones found at speeds 0.8 m/s, 1.2 m/s, and 1.6 m/s. It is clear that the method faithfully extrapolates the curves, by carrying curve shape and appearance, as speed increases, or decreases, where the extrema of the test points are shown with two sets of dashed curves. The curves compare well with the observed simulated gait-cycles in Fig. 4.18.

The prior expectation of our experiments is that the generated curves should fall within the same domain as the experimentally found curve. As we are not comparing to held-out data here (we will however in Section 5.4), we are merely interested in seeing how GPR compares to

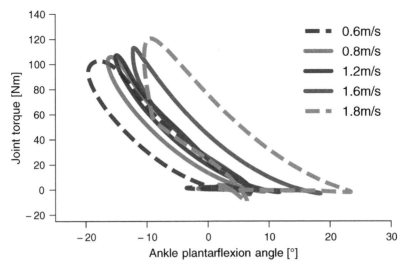

Figure 4.19 Measured and inferred ankle kinetics for subject 6. Curves are shown for velocities inside and outside the training data. Dashed curves lie on the extrema of the test points and the *experimental* values for the curves of 0.8 m/s, 1.2 m/s, and 1.6 m/s are shown as well. *(Adapted from [73]).*

other methods with respect to propagating the aforementioned curve properties.

In the following figures we compare GPR to common interpolation and extrapolations methods. We apply these methods to the three training velocities used thus far, for the ankle and knee moments and knee inflexion and ankle plantarflexion angles. Following this we interpolate and extrapolate the manifold up to and including 0.6 m/s and 1.8 m/s, in order to allow us to compare their regression properties to GPR (Fig. 4.20).

5.3.1 Nearest neighbor interpolation

In Fig. 4.21 the results for nearest neighbor interpolation have been superimposed on those of GPR for the same task. The GPR used $M_{52} - \text{ARD}$ and SE-ARD kernels for angles and moments respectively (throughout). As can be seen from both subplots, the method captures the lower bound well (0.6 m/s) but does not capture the correct shape of the upper range of velocity. *Why?* The algorithm selects the value of the nearest point but does not consider the values of other, neighboring points, yielding a piecewise-constant interpolant. For a shape as nonlinear as the torque-angle curve, this yields an unsatisfactory result. Second, this algorithm is primarily intended for unstructured inputs, hence why it is used in this "naive" comparison

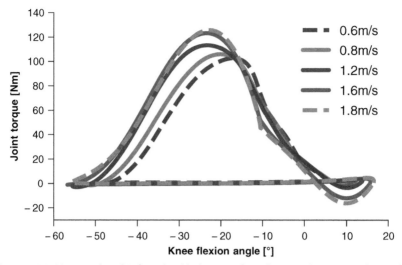

Figure 4.20 Measured and inferred ankle kinetics for subject 6. Curves are shown for velocities inside and outside the training data. Dashed curves lie on the extrema of the test points and the *experimental* values for the curves of 0.8 m/s, 1.2 m/s, and 1.6 m/s are shown as well. *(Adapted from [73]).*

fashion. Our data, though very sparse, is not unstructured when passed to the interpolant.

5.3.2 Linear regression

In Fig. 4.22 we employ a piecewise linear interpolant. We receive much the same results as for the nearest neighbors interpolant in Fig. 4.21. Again, it is not difficult to see why this produces poor results. We are trying to linearize a highly nonlinear surface, with a method poorly suited for that purpose. That being said, the benefits of these linear methods, are that they are simple, do not require any function selection (i.e. a kernel) and they are comparatively fast (this is not really an issue for us, but for larger problems it could be). Continuing, see that the solid lines in Fig. 4.22 do a poor job of regressing the surface to the extremities, worse in fact than the nearest neighbor interpolant. Granted, the performance is somewhat better for the knee curve in Fig. 4.22B than the ankle curves in Fig. 4.22A. In detail, the interpolant constructs a triangulation of the inputs using convex hull constructor (the convex hull in Euclidean space is the smallest convex set, that contains the set of points) and then, on each triangle, performs linear interpolation. This triangulation will be significantly warped when applied to semi-structured data such as ours, possibly why linear regression performs as badly as shown.

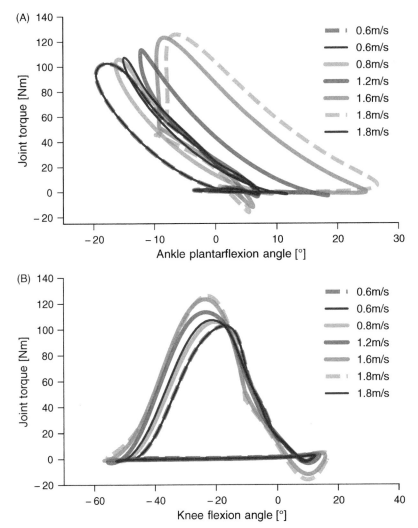

Figure 4.21 Nearest neighbor interpolation superimposed on the same regression task as GPR. (A) Ankle. (B) Knee. Shaded curves show the GPR results, and solid lucent curves, show the comparison method. For each manifold, training data was used at speeds 0.8m/s, 1.2m/s and 1.6m/s and GPR used to regress each variable in the envelope, to a speed domain of 0.6m/s to 1.8m/s, over one gait cycle. *Refer to Fig. 4.4 for the reference frame.*

5.3.3 Piecewise cubic curvature-minimizing interpolation

In Fig. 4.23 results are shown for cubic spline interpolation. Again, as this is a naïve method; we do not expect good results on this task. Because this method operates much as the nearest neighbor interpolant, with regards

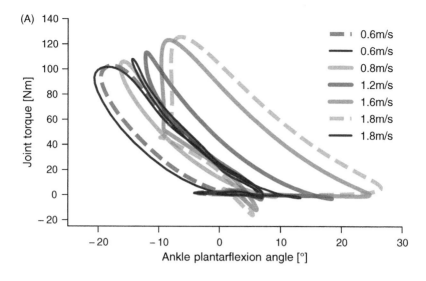

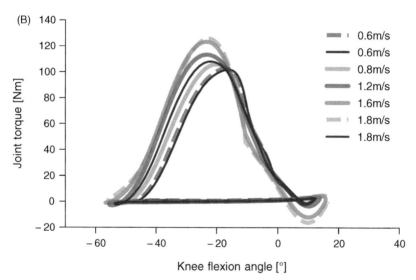

Figure 4.22 Linear regression superimposed on the same regression task as GPR. (A) Ankle. (B) Knee. Shaded curves show the GPR results, and solid lucent curves, show the comparison method. For each manifold, training data was used at speeds 0.8 m/s, 1.2 m/s, and 1.6 m/s and GPR used to regress each variable in the envelope, to a speed domain of 0.6–1.8 m/s, over one gait cycle. *Refer to Fig. 4.4 for the reference frame.*

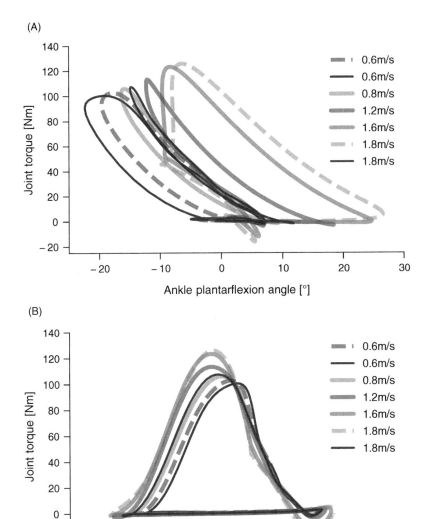

Figure 4.23 Cubic spline interpolation superimposed on the same regression task as GPR. (A) Ankle. (B) Knee. Shaded curves show the GPR results, and solid lucent curves, show the comparison method. For each manifold, training data was used at speeds 0.8 m/s, 1.2 m/s, and 1.6 m/s and GPR used to regress each variable in the envelope, to a speed domain of 0.6–1.8 m/s, over one gait cycle. *Refer to Fig. 4.6 for the reference frame.*

to the unstructured nature of the data, and the subsequent triangulation, a similarly poor result is achieved. Of the three methods, this is perhaps the worst, at least for the ankle regression, where the methods do not have a similar spread as the experimental data, nor GPR. Naturally we should not be comparing GPR and this method on a one-by-one basis, rather comparison is done by considering how well they place onto the experimentally obtained curves. We expect the solid lucent curve, in Fig. 4.23 to be aligned with, or above, the experimentally obtained counterpart at 1.6 m/s. This is not observed for the ankle. Rather, it is wedged in between 0.8 and 1.2 m/s, thus underestimating the torque-angle relationship.

Having employed naïve methods for comparing GPR, we now employ slightly more involved schemes for measuring, numerically, the performance of GPR on a held-out task.

5.4 Torque-angle relationship for held-out observations

In the study by Liu et al. [15], the authors studied muscle contributions related to providing vertical support and forward progression of the mass center of the human body. To quantify these contributions, over a range of walking speeds, three-dimensional muscle-actuated simulations of gait were generated and analyzed for eight subjects walking overground at very slow, slow, free, and fast speeds [15]. To examine the contributions of muscles to the acceleration of the mass center, they used gait analysis data at four walking speeds. We will use these observations in our held-out experiment, to measure the NLL predictive performance, using the posterior mean surface, compared to other involved methods.

What are involved methods? By "involved" we mean methods that bear some resemblance to GPs. In [43] and [58], authors note that there are a number of methods that can be interpreted as instances of GPs. Some include: generalized linear regression, neural networks (become GPs when there are infinite amount of hidden units), spline models, and support vector regression. We shall consider splines for their inherent simplicity.

All models are trained on three sets of curves, constituting the plantarflexion angle, recorded at three different velocities: 0.8 m/s, 1.17 m/s, and 1.64 m/s. The regression problem seeks to estimate, under these training examples ($N = 3$), the response at 0.61 m/s. These data points are selected from an 18-year-old female subject, weighing 63.1 kg. She was chosen because she bears the greatest physiological resemblance to the others subjects in this chapter (the other subjects in the cohort are all children).

Before looking at the results, lets consider some radial basis functions (RBF). Note that radial basis function interpolation is a common approach to scattered data interpolation [59]. The RBF and GPR models initially seem quite different. GPR is a weighted sum of the data, whereas RBF is a weighted sum of a kernel indexed by the distances between the data. Anjyo and Lewis [59] demonstrate that under some conditions they are in fact the same, in particular when one employs a Gaussian RBF, the GPR model is received. For this reason we shall not employ a Gaussian RBF, as we are more interested in properties of the RBF model, that do not yield a model equivalence.

We consider the multiquadric and cubic RBFs:

$$\varphi(r) = \sqrt{1 + (\varepsilon r)^2} \tag{4.14}$$

and

$$\varphi(r) = r^3 \tag{4.15}$$

where $r = \|x - x'\|$. Interpolation functions generated [60] from an RBF, are represented as

$$g(x) = \sum_{j=1}^{N} \alpha_j \varphi(\|x - x^j\|). \tag{4.16}$$

The goal is then to construct an estimation model to $f(x)$, that is, the latent underlying dynamics in our plantarflexion angle for example. We will not provide more detail on this model, but refer the interested reader to the references for further details.

The results from the regression task is pictorially shown with function heatmaps in Fig. 4.24 and numerical results are found in Table 4.5 as well as a method-by-method graphical comparison shown in Fig. 4.25.

With Fig. 4.24, Table 4.5, and Fig. 4.25, we are in a position to make some concluding remarks regarding these regression models. First, the RBF model is competitive when compared on a held-out data task such as this one. Indeed, the plantarflexion angle is tracked almost as well, as for with the GPR model. However, GPR does record a higher RMSE, especially using the Matern kernel, as it is a very smooth covariance function, which is a sound choice given the smooth nature of our test target, as well as our training data. Moreover, GPR gives us uncertainty bounds—as Fig. 4.25 shows. These uncertainties are numerically quantified in Table 4.4 as root mean-squared error (RMSE) scores. Alas, we demonstrate that GPR, for

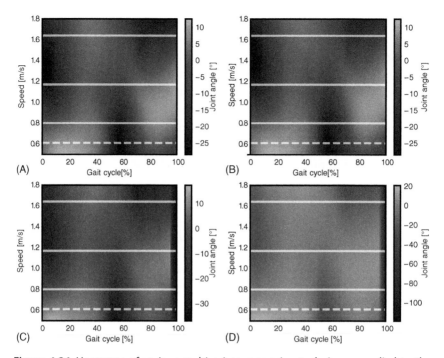

Figure 4.24 Heatmaps of various multivariate regression techniques, applied to the same problem. (A) SE-ARD kernel. (B) M_{52} – ARD kernel. (C) Multiquadric RBF. (D) Cubic RBF. The *top* row shows GPR with two different kernels. The *bottom* rows shows RBF interpolation with two different basis functions. The *solid horizontal white* lines correspond to the training observations **X** and the test-input X. is depicted by the *dashed horizontal white* line. The purpose of this exercise it to predict the plantarflexion angle response at the dashed line.

Table 4.5 Root mean squared error (RMSE) on held-out data from [15]. The related regression task for these results are shown in Fig. 4.24. The results from the GPR are the posterior predictive mean values.

Model	Heatmap	RMSE [°]
GPR with SE–ARD kernel	Fig. 4.24A	3.73 ± 1.48
GPR with M_{52} – ARD kernel	Fig. 4.24B	6.58 ± 0.17
RBF with multiquadric kernel	Fig. 4.24C	7.30
RBF with cubic kernel	Fig. 4.24D	9.51

this task, is superior to competitive methods on the same task. No doubt there will be other methods, that perform better, but most likely not at the same cost. The GPR model, for a small dataset like this, performs very well, on what is a difficult nonlinear manifold estimation task. Certainly, the RBF

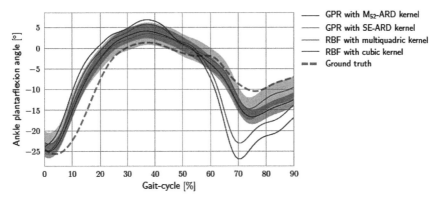

Figure 4.25 Shown are the results, for different methods, at inferring the ankle plantarflexion angle for subject 8, from the [15] dataset. Superimposed are the resulting curves from the respective regression manifolds shown in Fig. 4.24 (cross-section at the *dashed white* line is shown above). The ground-truth, that is, the experimentally measured values, are shown by the opaque dashed curve. Corresponding numerical RMSE results are shown in Table 4.4. The posterior uncertainty estimates for the GPR model are shown with opaque bounds on the posterior mean function for both model instances (i.e., different kernels).

model does well too, but with the drawback that this model does not yield uncertainty bounds.

5.5 Hardware experiments

In this section, the results of a healthy subject walking in a straight line at two different speeds (outside our training data) with the help of the ankle-foot prosthesis are presented. In this experiment, only the DP DoF of the ankle-foot prosthesis is used to provide push-off while IE is controlled to stay at zero. Fig. 4.26 illustrates the amount of push-off trajectory and power provided by the ankle-foot prosthesis at two different speeds. It is shown that in order to sustain the balance at higher walking speeds, the increased power at the ankle-foot prosthesis is necessary. A video^e of this experiment is also provided with this chapter to better illustrate the results. Current devices attempting the same task are typically equipped with controlled actuators which can replicate biomechanical characteristics of the human ankle (in as much as they have been tuned for that specific gait), improve the amputee gait and reduce the amount of metabolic energy consumed during locomotion. However, as also noted by Mai and

---

^{e} See the supplementary material at: https://youtu.be/FGnhBkR7xD0

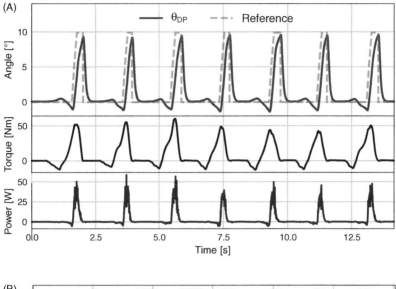

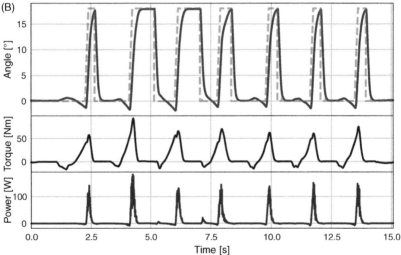

Figure 4.26 Experimental results of walking at two different speeds v_1 and v_2 with R^2 and RMSE measured for each trial. (A) $v_1 = 0.37\text{m/s}; \text{RMSE} = 2.54°, R^2 = 0.47$ (B) $v_2 = 0.6\text{m/s}, \text{RMSE} = 4.51°, R^2 = 0.67$. In the top plot the walking speed is v_1 and the injected push off power is well below 50W. In the bottom plot the walking speed is v_2 and the injected push off power is over 100W. *(Adapted from [73]).*

Commuri [61], the functioning of such devices on human subjects is difficult to test due to:

- changing gait;
- unknown ankle dynamics;

- complicated interaction between the foot and the ground;
- complicated interaction between the residual limb and the prosthesis.

We do not propose to deal with all of these raised issues. But we can deal with the first one. Alas, the observations $\mathbf{X}$ that we use for this experiment, are those extracted from the [17] dataset. Explicitly this means that for regression task, demonstrated in the previous simulation section; $N = 3$, which is to say for multivariate regression, we use three curves, evolving in time, of various features related to locomotion. These features are listed in Table 4.2 as well as the kernel combinations we use for each, as applied to the GPR model.

5.5.1 Preprocessing and single-cycle extraction

There are two main tasks we deal with in this section (1) preprocessing of the data and (2) single gait-cycle extraction. We start with preprocessing.

The data was recorded with a 100 Hz sampling rate, the dataset is cleaned such that missing values are interpolated using a variety of interpolation methods. Mostly though, the data was complete, and less than 0.5% were missing, hence little data in-painting was required. Following the procedure in [17] we compute gait events (toe-off and heel strike times), basic 2D inverse kinematics and dynamics, to get the actual endogenous angles from exogenous marker trajectories. This pre-processing protocol was constructed so that each gait event was stored in an array, each on a separate row, such that the mean gait-cycle could be calculated, with uncertainty bounds as shown in Fig. 4.5. We use the mean gait-cycle for all our experiments, but note that it is fully possible to incorporate the uncertainty bounds in our analysis as well. However, given that in this chapter, we are only interested in a proof-of-concept, we leave that exercise for future work.

Hence, in this set of experiments, we revert to being parsimonious with our data: we use mean quantities unless otherwise specified.

5.5.2 Results

The main results are shown in Fig. 4.26.

Our results contrast well with one of the studies which most similar to ours, namely those of [10,19,61]. Starting with ref. [10], they integrate biologically accurate torque-angle curves (two curves sampled at 0.5 m/s and 1.75 m/s) into their controller, by encoding "a few speed-specific curves from able-bodied studies" [10], and then interpolating (but *not* extrapolating) between them. The nature of their implementation means that they do not have uncertainty feedback incorporated into their controller.

Nonetheless, while comparison is difficult (owing to the different nature of our experiments), similar root mean squared error (RMSE) and coefficient of determination (R^2) scores are recorded. We measure the metrics between the experimental values and the simulated values, for the same experiment. Joint angles are computed from the mathematical model in simulation, thus allowing us to track the error. Returning to [10], for their mid-stance to late-stance, they recorded an RMSE of 3.05 degree and R^2 of 0.7154 when subjects walked "on a treadmill at continuously varying walking speeds" [10]. Their reference is derived from able-bodied subjects, not a simulated one, so our R^2 are not quite comparable. Moreover, they use the full manifold regressed, whereas we are only using the push-off part in our simulations, to elicit realistic power, torque and angle profiles.

Another study, which bears comparing against, is the seminal study by Markowitz et al. [19]. They use a muscle-tendon model, which, conditioned on observations, produces estimates of the activation, force, length, and velocity of the main muscles spanning the ankle. These are used to derive control feedback loops that may (from a neural perspective) be critical in the control of those muscles during walking. This allows them to closely reproduce the muscle dynamics estimated from biological data. They produced similar results, to those we display in Fig. 4.19. A rather simplistic result, but nonetheless one which suggest that our method is salient and concurrent with other state-of-the-art methods. 'Similar' here means that like us, they carry shape, size and placement of the torque-angle curves shown in Fig. 4.19. Which is to say, their method supports their hypothesis, as does ours: the curves should stray away from the center with an increase or decrease in speed. While we naturally cannot confirm that our model forms the underlying mechanism for this behavior, empirical evidence suggest that it supports that thesis. Further, they present results which demonstrate the prosthesis ankle and knee angles and torques, measured during their clinical trials, against those from a height and weight -matched healthy subject. Though they do not provide RMSE or R^2 scores, inspection suggests that our method is comparable in speed adaptability. That being said, because we propose a probabilistic method, the variance on our predictions *may* reduce as the size of **X** grows, given that new samples increase our knowledge of the underlying state-space of the latent manifold which we are trying to estimate with our GPR model. This is not true for their comparable controller, which, although it is a tunable system, is not nonparametric.

Finally and more recently, a study was presented by Mai and Commuri [61], which shows very impressive results. Like ourselves, [19] and [10], they study transtibial (amputation between knee and ankle joint) ampu-tees. They present an artificial neural network-based hierarchical controller that recognizes the amputees' intent from measured gait observations. Once this is done the controller selects a displacement profile for the prosthetic joint based on the amputees' intent, and then adaptively "compensates for the un-modeled dynamics and disturbances for closed-loop stability with guaranteed tracking performance" [61]. Effectively the result of their approach used gait-based quantities, collected from a group of nine trans-tibial amputees, to calculate an appropriate control torque for the recog-nized gait. Regrettably their results are only simulated, using their collected data. However their ankle plantarflexion angle tracking is impressive and comparable to our own results. That being said, while their tracking perfor-mance is impressive, their transition tracking could be improved see ([61], Fig. 8). Not withstanding they present two challenging scenarios: one with artificial noise added, and one involving velocity changes to the profile that their controller is trying to track. This was achieved by a neural network to learn the nonlinear ankle dynamics, and the interaction between the foot and the walking terrain was compensated by an empirical model of the ground reaction force [61]. To this end they receive a control torque which they apply to the prosthetic ankle joint to track the reference ankle displacement during gait. Now, earlier on we noted that their results were comparable to our own. This is under the proviso of merely estimating the error from their plots, as they have not presented RMSE or NLL scores of their method. However, as their goal is virtually identical to our own, they manage to track their reference trajectory 'better' than ourselves (still; only in simulation), in as much as the error can be estimated from their plots.

Finally, while we present competitive results to our peers, there are drawbacks to our method. While Gaussian process regression is attractive for its many aforementioned properties, it also contains many drawbacks, and the way in which we have used the model, may not be realistic. Lets start with our primary assumption in this work: *gait-cycles are atomic*. This means that we posit that we can construct gait-cycles of arbitrary lengths, whose coordinates, be it exogenous or endogenous, we can pass as reference trajectories to a controller, which can then be tracked to elicit locomotion. The first form of criticism that can be leveled at this assumption, is that it could be a incorrect modeling assumption. Locomotion, from the body's point of view, could be a long-term planning process, which does not con-stitute atomic features of the nature that we propose. Indeed, we posit that the human onboard control system is Bayesian in that it stores local model of motion, such as the ones we demonstrate in this chapter, and deploys them according to user intent. For example, if the user wants to run 100 m then the biological control system, adapts, online, to create gait-cycles that correspond to this user intent, with bounds placed by biomechanics, agility and resources (i.e., available energy). An alternate strategy suggests that the control system plans well in advance, the whole 100 m stretch, in intricate detail, and the slightly adapts this principle during the run, with minor adjustments throughout.

Second, we have not covered *how* we propose to combine atomic gait-cycles to elicit natural looking locomotion. While we have combined atoms to render longer sequences, these were merely concatenated, and where there was missing data, points were interpolated. This, as it turns out, was an adequate strategy, which did not present any significant problems, neither methodologically nor in the end result. Realistic however, it would be fair to suggest that were this model deployed on a real, commercial, prosthesis, procedures would be required for interpolating between atoms as they are deployed, based on user intent. This is not to say that it is not possible, it is, as we have demonstrated. It is rather to emphasize that we have yet to conduct incisive experiments into this area, hence it would be hard deign at this point what the optimal strategy is for concatenating atomic gait features online (or indeed offline).

How do humans transition? There are preciously few studies that actually consider *how* we move from one velocity to another, to take the simplest of examples. Consider a complex one instead: how do we go from walking to climbing up a staircase—what is the dynamic response, once intent has been established? In this study, for simplicity, we assumed that the transition was linear. Hence, in our transition experiments above, we simply apply a linear transition from one velocity to another. It is highly improbable that this is the actual mechanism employed by the human body. This is simply because human motion is nonlinear, highly nonlinear, and thus it is also plausible to suggest that so too is human activity transition.

6 Discussion and conclusion

We have demonstrated the utility of using GP regression for finding what we term a *locomotion envelope*, which can serve to synthesize a high variety of locomotion for a powered ankle-foot prosthesis. There is a need to develop a control strategy which can provide biologically sound torques across a wide range of walking speeds without requiring velocity-specific control tuning.

The benefits of our approach, in particular, is its offline construction, thus making it fast and robust (this does not mean that it cannot adapt online, but in its present incarnation it does not have to). More importantly, since we use supervised learning methods, we directly employ human demonstrations, hence achieving prosthesis locomotion that is natural looking (though we must stress that this is only yet at the simulation stage), and numerically consistent with experimental human locomotion – see Fig. 4.19.

There are a number of issues to consider henceforth. First, while it is useful to have the capacity to arbitrarily switch and transition between velocities, we have not discussed intention estimation, or a specific perception layer of our method. This was deliberate but it bears considering, given the medical nature of our chosen domain. First, in future work, in a clinical setting, we propose to place markers, in the sagittal plane, of the ankle and knee of the subject in order to measure their speed. Depending on which range they fall into, an appropriate speed will be selected by the controller. This, of course, is easy in such a controlled environment. We require high-level controls [5], of which there are many, such as decision trees and finite state machines. These methods, albeit simple, are robust and operate on a set of identified rules, which dictate when a mode-transition takes place. As input they could take the user-state (e.g., acceleration measurements from the healthy leg) or environmental queues such as frictional response of the walking surface. There are also promising methods to be found in more classical machine learning classifiers such as Gaussian mixture models [62] or support vector machines [63].

For Gaussian processes, the limiting factor with long-range predictions for all the manifolds, is the mean function posterior predictive uncertainties. We typically use a zero mean function, but there is ample evidence to suggest that using a more domain-specific mean function will allow us to make robust long-range predictions (i.e., inferring the manifold shape, magnitude and temporal evolution at speeds far away from the training data $s_\star^- \ll s \ll s_\star^+$ where s are the speeds present in the design matrix $\mathbf{X}$). Moreover, as we have already alluded to; kernel tuning is still required. This is a drawback, but necessary to incorporate our prior domain knowledge into the predictive framework. But we are rewarded with a smaller variance in our prediction should we tune the kernel accurately. This is something that should not be understated. The size of our variance implicitly gives us a *confidence level* of our prediction. This means that we can assign appropriate action to prediction results, depending on how certain we are of them. Equally, we can determine if our training database is sufficiently and appropriately broad for the range of motion, which we want the prosthesis to be able to undertake [29]. Though we only discuss Gaussian processes in the context of nonlinear regression, there are many contenders for this role, such as nonlinear least squares regression, neural networks and support vector regression.

The second limiting factor that we mentioned at the start of the previous paragraph, is the predictive uncertainty— given our model, how sure can we be that its predictions are accurate? [64] discuss this problem in detail. We employ Bayesian probability to perform inference about quantities which are unknown to us, or which we seek to increase our knowledge of. The GP paradigm is a good tool for achieving such information exploration. But even through principled model design, and considerable prior knowledge, we cannot avoid uncertainty. [64] and [43] relate that this uncertainty can take many forms. For example, in real experiments data is typically: missing as a results of sensor failure; multiple sensors will often be correlated; there may be a complex noise which cannot be assuaged by assuming a simple additive noise model and many others. All of these items play into our predictive uncertainty, which we can regularize in a few ways: by collecting more observations, to overcome sensor failure and eventually to overwhelm an mispecified prior (e.g., picking the wrong kernel). Better sensors are always desirable, but this is more subject to resources, than any explicit fault of the theory.

These are but general aspects of GP regression, for our setup we performed each surface-learning in isolation which is to say that no information was passed between manifolds during regression – they were all independent. We posit that by jointly learning the envelope, with a regression taking place over all constituent variables, with a global envelope training target, we would have received better posterior predictive performance. This could in the future be tackled by employing twin GPs (TGP) – a generic structured prediction method that uses GP priors on both covariates and responses. The TGP method models interdependencies between covariates, as in a typical GP, but also those between responses, hence correlations among both inputs and outputs are accounted for [65]. More promising still is the notion of coregionalized Gaussian processes [66], as an extension to our locomotion envelopes.

In addition, it is possible, using neural networks, to learn the whole locomotion envelope in one go – see the work by [67]. Like other kernel methods, GPs are useful because they have a covariance functions whose hyperparameters can be learned by maximizing the marginal likelihood

[68] (as ever though, there is risk of overfitting). Given the values of their hyper-parameters, and this often allows a fine and precise trade-off between fitting the data and smoothing. While we are using a comparatively small number of data points to describe our variables over one-gait cycle, this is still a computationally viable approach. But one could imagine a scenario where there is a need to regress over several gait-cycles (to capture a long-term dependency, e.g. of a subject's particular gait), then the method needs to scale to accommodate this larger dataset. This, GPs cannot do well, yet.

As noted by [43], [69], and [70], the generic inference and learning algorithms for GPR has a runtime of $\mathcal{O}(N^3)$ and $\mathcal{O}(N^2)$ memory complexity when N is the number of observations in $\mathbf{X}$. To be a bit more specific, prediction cost per test case is $\mathcal{O}(N^3)$ for the mean and $\mathcal{O}(N^2)$ for the variance [58], [43]. This penalty results from using the Cholesky decomposition for finding the inverse of the covariance matrix. Recent work by, for example, [71] and [70] demonstrate that sparse approximations to the full GPR problem, allow the latter to scale without unreasonable time and space penalties. Looking further afield there are alternatives, such as the modern version of neural-nets; deep-learning, which could also prove useful in the setting of nonlinear regression for manifold learning. The obvious drawback with these methods, as mentioned earlier, is that they will not yield uncertainty bounds on the predictions. We have not used our uncertainty bounds in this chapter, but we have the option of doing so, whereas a neural network for example, has a more involved methodology for producing error bars on the predictions.

Finally, we found information lacking in the precise nature of gait-cycle transition i.e. the way by which locomotion transitions occur between gait-cycles at different speeds. Whilst most current control methods divide the gait cycle into several sequential periods, each with independent controllers. It is not clear how precisely current switching modalities mimic those exhibited by human locomotion. Whilst we have taken the view in this study that there is large acceleration at the beginning of the gait-cycle (see Fig. 4.13) required to accurately mimic human transition, Van der Noot et al. [35] enforce a fixed time for each transitions in between speeds. Overall we have found the literature in this area wanting, thus revealing a need for further studies in this domain.

6.1 Conclusion

In this chapter, we have presented a data-driven control strategy for ankle-foot prostheses. We have demonstrated (by way of simulation and initial hardware tests) that the methodology has the capacity to allow the user to walk over a wide range of speeds, while also providing for fast variations outside of the training data. In future work, we intend to expand upon the speed range by adopting a more domain-specific mean function for GP regression as well as employing more advanced forms of information sharing in the GP framework.

Though this work is primarily intended for the rehabilitation robotics domain, it may prove insightful to the field of bipedal humanoids. Although we have implemented the work for prostheses, the underlying theory concerns basic understanding of human locomotion, and how to adapt those insights to generalize our control strategies, to effect a singular, or desired set of locomotions. Further, we have demonstrated that taking a broadly cyclical view of human locomotion is useful since it means we can extrapolate over gait–cycles rather than full time-series data, across speeds. More importantly we have demonstrated a method for generating biologically *plausible* torque–angle curves.

As ever though, there is much room for improvement, and we have mentioned a few already. A large problem which we came upon several times during this work, was the difficulty in comparing with other powered prostheses control strategies, thus making it more difficult to quantitatively compare our computational simulations to those of our peers.

This was not merely an issue of implementing their methods. Because almost all studies in this area employ some form of training data (which is sensible, given that we are trying to mimic human physiological output), which they use for their controllers, was difficult to obtain. Most likely because it was expensive to obtain and is bound by institution-specific rules and regulations. However, the datasets that we do use in this study [15–17] are not specifically designed for prostheses controller design. They are more concerned with muscle force generation, and modeling how the human body actually does this. This is a different goal to our own, where we are trying to artificially create that force generation and anthropomorphically pass it to the end effectors of our device. Certainly though, our goals are very much aligned, but an optimal dataset for our purposes would involve, for example, multiple subjects walking at multiple speeds (> 10), for several hours each. This would no doubt be expensive to collect.

Moreover, there are issues with scaling these datasets. Because we suppose that there are latent, common dynamics, to human gait, the observations that we do receive have to be appropriately scaled—a plantarflexion angle for a child will be different to that of an adult. There is no standard protocol to receive nondimensional metrics of curves, and many studies employ their own for various quantities. For example, Liu et al. [15] scales the velocities in their study by $\sqrt{gL_{leg}}$ where g is gravitational acceleration and L_{leg} is the measured length of a subject's leg. This is good approach to nondimensionality, however not one that can be carried across to the study by Moore et al. [17], since they do not measure the length of participants' legs. Consequently, as there is as of yet no large body of work concerning control schemes for speed-variations for prostheses (granted, a rather niche topic), common metrics will have to designed for *what* data needs to be collected for robust control design, and *how* it is collected so that it can easily be *shared* with the wider community. For now though, to quote [17]:

Even though years of data on thousands of subjects now exist, this data is not widely disseminated, well organized, nor available with few or no restrictions

However, as the area matures, especially with bipedal robots, datasets will surely (we say with optimistic hesitation) become more widely available.

References

[1] B.D. Argall, Machine learning for shared control with assistive machines. in: ICRA Workshop on autonomous learning: from machine learning to learning in real-world autonomous systems. ICRA, 2013.

[2] J. Morimoto, T. Noda, and S. Hyon, Extraction of latent kinematic relationships between human users and assistive robots. in: Robotics and Automation (ICRA), IEEE, (2012) 3909–3915.

[3] C.A. Cheng, T.H. Huang, H.P. Huang, Bayesian human intention estimator for exoskeleton system, in: Advanced Intelligent Mechatronics (AIM), IEEE/ASME, IEEE, (2013) 465–470.

[4] E. Aertbeliën, J. De Schutter, Learning a predictive model of human gait for the control of a lower-limb exoskeleton. in International Conference on Biomedical Robotics and Biomechatronics, 2014.

[5] M.R. Tucker, J. Olivier, A. Pagel, H. Bleuler, M. Bouri, O. Lambercy, J.d.R. Milln, R. Riener, H. Vallery, R. Gassert, Control strategies for active lower extremity prosthetics and orthotics: A review, J. NeuroEng. Rehab. 120 (1) (2015) 1.

[6] D. Grimes, W. Flowers, M. Donath, Feasibility of an active control scheme for above knee prostheses, J. Biomech. Eng. 990 (4) (1977) 215–221.

[7] W.C. Flowers, R.W. Mann, An electrohydraulic knee-torque controller for a prosthesis simulator, J. Biomech. Eng. 990 (1) (1977) 3–8.

[8] J.L. Stein, W.C. Flowers, Stance phase control of above-knee prostheses: Knee control versus sach foot design, J. Biomech. 200 (1) (1987) 19–28.

[9] C. Ferreira, L.P. Reis, C.P. Santos, Review of control strategies for lower limb prostheses, in Robot 2015: Iberian Robotics Conference, Springer, (2016) 209–220.

[10] T. Lenzi, L. Hargrove, J. Sensinger, Speed-adaptation mechanism: Robotic prostheses can actively regulate joint torque, Rob. Automat. Mag. IEEE 210 (4) (2014) 94–107.

[11] J.K. Hitt, T.G. Sugar, M. Holgate, R. Bellman, An active foot-ankle prosthesis with biomechanical energy regeneration, J. Med. Dev.-Trans. Asme 40 (1) (2010).

[12] F. Sup, A. Bohara, M. Goldfarb, Design and control of a powered transfemoral prosthesis, Int. J. Rob. Res. 27 (2008) 263–273.

[13] F. Sup, H.A. Varol, M. Goldfarb, Upslope walking with a powered knee and ankle prosthesis: Initial results with an amputee subject, Neur. Sys. Rehab. Eng. IEEE 190 (1) (2011) 71–78.

[14] F. Sup, H.A. Varol, J. Mitchell, T.J. Withrow, M. Goldfarb, Preliminary evaluations of a self-contained anthropomorphic transfemoral prosthesis, IEEE/ASME Trans. Mechat. 140 (6) (2009) 667–676.

[15] M.Q. Liu, F.C. Anderson, M.H. Schwartz, S.L. Delp, Muscle contributions to support and progression over a range of walking speeds, J. Biomech. 410 (15) (2008) 3243–3252.

[16] E.M. Arnold, S.R. Hamner, A. Seth, M. Millard, S.L. Delp, How muscle fiber lengths and velocities affect muscle force generation as humans walk and run at different speeds, J. Exp. Biol. 2160 (11) (2013) 2150–2160.

[17] J.K. Moore, S.K. Hnat, A.J. van den Bogert, An elaborate data set on human gait and the effect of mechanical perturbations, Peer J. 3 (2015) e918.

[18] S. Au, M. Berniker, H. Herr, Powered ankle-foot prosthesis to assist level-ground and stair-descent gaits, Neur. Net. 210 (4) (2008) 654–666.

[19] J. Markowitz, P. Krishnaswamy, M.F. Eilenberg, K. Endo, C. Barnhart, H. Herr, Speed adaptation in a powered transtibial prosthesis controlled with a neuromuscular model, Philos. Trans. Royal Soc. London B: Biol. Sci. 3660 (1570) (2011) 1621–1631.

[20] L.J. Hargrove, A.M. Simon, A.J. Young, R.D. Lipschutz, S.B. Finucane, D.G. Smith, T.A. Kuiken, Robotic leg control with emg decoding in an amputee with nerve transfers, N Engl. J. Med. 3690 (13) (2013) 1237–1242.

[21] T. Lenzi, L. Hargrove, J. Sensinger, Speed-adaptation mechanism: Robotic prostheses can actively regulate joint torque, IEEE Rob. Autom. Mag. 210 (4) (2014) 94–107.

[22] D. Quintero, D.J. Villarreal, R.D. Gregg, Preliminary experiments with a unified controller for a powered knee-ankle prosthetic leg across walking speeds, in: 2016 IEEE/RSJ International Conference on Intelligent Robots and Systems (IROS), (2016) 5427–5433.

[23] H.M. Herr, A.M. Grabowski, Bionic ankle–foot prosthesis normalizes walking gait for persons with leg amputation. in: Proc. R. Soc. B, 279, pp. 457–464. The Royal Society, 2012.

[24] M.A. Holgate, A.W. Bohler, T.G. Suga, Control algorithms for ankle robots: A reflection on the state-of-the-art and presentation of two novel algorithms, in: Biomedical Robotics and Biomechatronics, IEEE, pp. 97–102. IEEE, 2008.

[25] R. Jimnez-Fabin, O. Verlinden, Review of control algorithms for robotic ankle systems in lower-limb orthoses, prostheses, and exoskeletons, Med. Eng. Phys. 340 (4) (2012) 397–408.

[26] J. Hong, C. Chun, S. J. Kim, Gaussian process gait trajectory learning and generation of collision-free motion for assist-as-needed rehabilitation, in: Humanoid Robots (Humanoids), IEEE-RAS, IEEE, (2015) 181–186.

[27] J.M. Wang, Gaussian process dynamical models for human motion. Master's thesis, University of Toronto, (2005).

[28] D.J. Lizotte, T. Wang, M.H. Bowling, D. Schuurmans, Automatic gait optimization with gaussian process regression, i IJCAI, 7 (2007) 944–949.

[29] Y. Yun, H.-C. Kim, S.Y. Shin, J. Lee, A.D. Deshpande, C. Kim, Statistical method for prediction of gait kinematics with gaussian process regression, J. Biomech. 470 (1) (2014) 186–192.

[30] S. Song, H. Geyer, Regulating speed in a neuromuscular human running model, in: Humanoid Robots (Humanoids), IEEE-RAS, IEEE, (2015) 217–222.

[31] A.J. Ijspeert, Central pattern generators for locomotion control in animals and robots: A review, Neur. Net. 210 (4) (2008) 642–653.

[32] N. Thatte, H. Geyer, Towards local reflexive control of a powered transfemoral prosthesis for robust amputee push and trip recovery, in: Intelligent Robots and Systems (IROS), IEEE/RSJ, IEEE, (2014) 2069–2074.

[33] M.F. Eilenberg, H. Geyer, H. Herr, Control of a powered anklefoot prosthesis based on a neuromuscular model, IEEE Trans. Neur. Sys. Rehab. Eng. 180 (2) (2010) 164–173.

[34] H. Geyer, H. Herr, A muscle-reflex model that encodes principles of legged mechanics produces human walking dynamics and muscle activities, IEEE Trans. Neur. Sys. Rehab. Eng. 180 (3) (2010) 263–273.

[35] N. Van der Noot, A.J. Ijspeert, R. Ronsse, Biped gait controller for large speed variations, combining reflexes and a central pattern generator in a neuromuscular model, in: Robotics and Automation (ICRA), IEEE, IEEE, (2015) 6267–6274.

[36] M. Vukobratovic´, B. Borovac, Zero-moment pointthirty five years of its life, Int. J. Human. Robot. 10 (1) (2004) 157–173.

[37] S.H. Collins, Dynamic walking principles applied to human gait. PhD thesis, University of Michigan, 2008.

[38] H. Gritli, S. Belghith, N. Khraief, Ogy-based control of chaos in semi-passive dynamic walking of a torso-driven biped robot, Nonlinear Dyn. 790 (2) (2015) 1363–1384.

[39] F. Iida, R. Tedrake, Minimalistic control of biped walking in rough terrain, Autonom. Robot. 280 (3) (2010) 355–368.

[40] P. Fisette, J. Samin, Robotran: Symbolic generation of multi-body system dynamic equations, in: Advanced Multibody System Dynamics, Springer, (1993) 373–378.

[41] D.M. Wolpert, Z. Ghahramani, Computational principles of movement neuroscience, Nat. Neurosci. 30 (11s) (2000) 1212.

[42] N. Dounskaia, Control of human limb movements: the leading joint hypothesis and its practical applications, Exer. Sport Sci. Rev. 380 (4) (2010) 201.

[43] C.E. Rasmussen, C.K.I. Williams, Gaussian processes for machine learning, The MIT Press, (2006).

[44] A. McHutchon, C.E. Rasmussen, Gaussian process training with input noise, in: Advances in Neural Information Processing Systems, (2011) 1341–1349,.

[45] A.M. Woodward, B.K. Alsberg, D.B. Kell, The effect of heteroscedastic noise on the chemometric modelling of frequency domain data, Chemomet. Intel. Lab. Sys. 400 (1) (1998) 101–107.

[46] P.W. Goldberg, C.K. Williams, C.M. Bishop, Regression with input-dependent noise: A gaussian process treatment, in: Advances in neural information processing systems, (1998) 493–499.

[47] E. Snelson, Z. Ghahramani, C.E. Rasmussen, Warped gaussian processes, in: Advances in neural information processing systems, (2004) 337–344.

[48] C.E. Rasmussen, Z. Ghahramani,. Infinite mixtures of gaussian process experts, in: Advances in neural information processing systems, (2002) 881–888.

[49] D.J. Farris, G.S. Sawicki, The mechanics and energetics of human walking and running: A joint level perspective, J. Royal Soc. Interface (2011).

[50] N. Hogan, S.P. Buerger, Impedance and interaction control, robotics and automation handbook, 2005.

[51] S. Chan, B. Yao, W. Gao, M. Cheng, Robust impedance control of robot manipulators, Int. J. Robot. Autom. 60 (4) (1991) 220–227.

[52] M.A. Mistry, Tutorial on impedance control and physical human-robot interaction. Available from: http://www.robot-manipulation.uk/impedance_control_tutorial.pdf, July 2017. Presented at the second UK Robot Manipulation Workshop.

[53] N. Hogan, Impedance control: An approach to manipulation, in: American Control Conference, IEEE, (1984) 304–313.

[54] Q. C. Pham, Examples: hybrid control and impedance control, 2016. Available from: http://osrobotics.org/pages/examples_force_control.html.

[55] E.M. Ficanha, G.A. Ribeiro, H. Dallali, M. Rastgaar, Design and preliminary evaluation of a two dofs cable-driven ankle–foot prosthesis with active dorsiflexion–plantarflexion and inversion–eversion, Front. Bioeng. Biotech. 4 (2016).

[56] D.K. Duvenaud, J.R. Lloyd, R.B. Grosse, J.B. Tenenbaum, Z. Ghahramani, Structure discovery in nonparametric regression through compositional kernel search, in: ICML 3 (2013) 1166–1174.

[57] D. Duvenaud, Automatic model construction with Gaussian processes. PhD thesis, University of Cambridge, 2014.

[58] E. Snelson, Tutorial: Gaussian process models for machine learning, Gatsby Comp. Neurosci. Unit, UCL (2006).

[59] K. Anjyo, J. Lewis, Rbf interpolation and gaussian process regression through an rkhs formulation, J. Math. Ind. 30 (6) (2011) 63–71.

[60] H. Rocha, On the selection of the most adequate radial basis function, Appl. Math. Model. 330 (3) (2009) 1573–1583.

[61] A. Mai, S. Commuri, Intelligent control of a prosthetic ankle joint using gait recognition, Cont. Eng. Prac. 49 (2016) 1–13.

[62] H.A. Varol, F. Sup, M. Goldfarb, Multiclass real-time intent recognition of a powered lower limb prosthesis, IEEE Trans. Biomed. Eng. 570 (3) (2010) 542–551.

[63] A. Kilicarslan, S. Prasad, R.G. Grossman, J.L. Contreras-Vidal, High accuracy decoding of user intentions using eeg to control a lower-body exoskeleton, in Engineering in medicine and biology society (EMBC), IEEE, IEEE, (2013) 5606-5609.

[64] M. Osborne, S.J. Roberts, Gaussian processes for prediction. Technical Report PARG-07-01, 2007.

[65] L. Bo, C. Sminchisescu, Twin gaussian processes for structured prediction, Int. J. Comp. Vis. 870 (1–2) (2010) 28–52.

[66] M.A. Alvarez, L. Rosasco, N.D. Lawrence, Kernels for vector-valued functions: A review, Found. Trend. Mach. Learn. 40 (3) (2012) 195–266.

[67] D. Holden, J. Saito, T. Komura, T. Joyce, Learning motion manifolds with convolutional autoencoders, in SIGGRAPH Asia Technical Briefs, ACM, (2015) 18.

[68] D. Barber, Bayesian Reasoning and Machine Learning, Cambridge University Press, (2012).

[69] M.P. Deisenroth, J.W. Ng, Distributed gaussian processes. arXiv preprint arXiv:1502.02843, 2015.

[70] Y. Saatci, Scalable inference for structured Gaussian process models. PhD thesis, University of Cambridge, 2012.

[71] A.G. Wilson, H. Nickisch, Kernel interpolation for scalable structured gaussian processes (kiss-gp). *arXiv preprint arXiv:1503.01057*, 2015.

[72] N. Dhir, F. Wood, Improved activity recognition via Kalman smoothing and multiclass linear discriminant analysis. in Proceedings of the Engineering in Medicine and Biology Society (EMBC), IEEE, IEEE, (2014) 582-585.

[73] N. Dhir, H. Dallali, E.M. Ficanha, G.A. Ribeiro, M. Rastgaar, Locomotion envelopes for adaptive control of powered ankle prostheses, in: Robotics and Automation (ICRA). IEEE, 2018.

CHAPTER 5

Disturbance observer applications in rehabilitation robotics: an overview

Alireza Mohammadi[a], Houman Dallali[b]
[a]Department of Electrical and Computer Engineering, University of Michigan, Dearborn, MI, United States
[b]Department of Computer Science, California State University, Channel Islands, Camarillo, CA, United States

1 Introduction

Using robotic devices for training during post-disability rehabilitation, where the objective is to restore the patient's sensorimotor functioning, has been on the rise in the last decade [1–3]. Numerous wearable exoskeleton robots have been designed for rehabilitation of patients' upper and lower extremities [4,5], spine [6], knee [7], ankle [8], and hand [9,10], among others. These robots are designed to provide human performance augmentation for rehabilitation of the targeted limbs after disabling events such as stroke. Robot-aided rehabilitation can significantly improve the state-of-the-art therapeutic and performance assessment procedures in the patient rehabilitation, thanks to their potential for relieving the time and effort burden on the clinicians, their high degree of repeatability, allowing more intensive limb trainings, and providing patients with more effective and accurate rehabilitation exercises.

In addition to the challenges that are associated with the mechanical design of rehabilitation robots [11–15], there are still many open questions regarding the development of effective control systems for such devices [16–19]. These questions include but are not limited to improving the physio-cognitive interaction between the patient and the rehabilitation robotic device, generalization of control strategies across various types of rehabilitation robotic devices, and proper control of the robotic device in concert with the patient's sensory-motor control system.

In rehabilitation robotics, the human and the robot are constantly interacting with each other. Consequently, measuring the interaction forces between the robotic device and the patient becomes of crucial importance to both the safety of the user and the efficacy of the rehabilitation robot

Powered Prostheses
http://dx.doi.org/10.1016/B978-0-12-817450-0.00005-5

control system. Furthermore, the need for adaptation and planning of limb training procedures [20,21] and assessing the patient's motor function in robot-aided therapy protocols [22,23] requires knowing the interaction forces between the patient and the robotic exoskeleton.

In rehabilitation robotic applications, there are certain drawbacks associated with inclusion of force sensors such as introduction of control system stability issues, narrow bandwidth of the force sensors, increase in the rehabilitation system cost, and the requirement for frequent calibrations by the therapists. Furthermore, in certain rehabilitation robotic applications such as hand exoskeletons, limited space makes replacement of force sensors by alternative measurement and estimation methods necessary.

A recent cost-effective alternative to inclusion of force sensors in the rehabilitation robot control systems is based on "sensorless force estimation," whereby the interaction forces between the robot and the patient are estimated using "observers." Observers, which can be considered as a replacement for force sensors, belong to a special class of estimation algorithms that use the dynamical model of the robot along with the robot joint position/velocity measurements to estimate the unknown inputs/forces exerted to the robot. The observers use the differences between the expected and the actual robot configurations to approximate the unknown input forces that are acting on the robot.

There are various methods for designing unknown input observers in robotic applications in the literature, whose complexity levels vary across their underlying estimation techniques such as Kalman filters [24,25], generalized momentum observers [26,27], and disturbance observers [28–32]. Among the aforementioned sensorless force estimation techniques, disturbance observers have found numerous applications in rehabilitation robotics.

Disturbance observers, which were introduced in the early 1980s for robust control in mechatronics [33–36], have been employed across various fault detection and robust control applications. The role of disturbance observers is to reconstruct, from the measured output variables and the known inputs applied to the system, unknown disturbances/inputs that are acting on the control system. The output of the disturbance observer, which provides an estimate of the unknown inputs, can then be employed for various purposes such as fault detection or feedforward compensation of disturbances that are acting on the control system.

According to the generalized control framework in Refs. [37,38], where the control approaches in rehabilitation robotics are classified, a hierarchical control structure should be adopted for these robotic devices. The hierarchical controller at the high level should be capable of perceiving the patient's intent during locomotion and/or object manipulation as well as issuing high-level motor commands and volitional control of the robotic device. In the mid-level, where the high-level commands are integrated with feedback from sensors, the patient's intentions should be mapped to desired device state outputs. Finally, in the low- level, the robotic device actuators are controlled by means of control schemes such as torque, impedance, position/velocity, or admittance controllers to track the desired commands from the mid-level controller. As it will be seen in this review, disturbance observers play an important role in all of the aforementioned layers of hierarchy of rehabilitation robot control systems.

In this work, a classification and applications of disturbance observers in rehabilitation robotics are presented. To this end, the applications of disturbance observers in disturbance and force estimation in upper-body rehabilitation, lower- body rehabilitation, and telerehabilitation are studied. This study demonstrates that employing disturbance observers in rehabilitation robotic applications can be beneficial for simplifying the complexities that are associated with employing expensive force sensors in rehabilitation robotic devices. Furthermore, the plethora of applications of disturbance observers in robotic rehabilitation opens a wide research area for using the disturbance observers in this field.

This overview is performed in three main areas of rehabilitation: (1) upper-body rehabilitation; (2) lower-body rehabilitation; and (3) telerehabilitation. The rest of this chapter is organized as follows. First, in Section 1.2, preliminaries on disturbance observers are briefly reviewed. Next, in Section 1.3, the applications of disturbance observers for upper-body rehabilitation are presented. Thereafter, in Section 1.4, an overview of applications of disturbance observers in lower-body rehabilitation is provided. Afterward, in Section 1.5, the specific issues associated with robotic telerehabilitation along with applications of disturbance observers for overcoming these issues are discussed. Finally, the concluding remarks are presented in Section 1.6.

2 Background on disturbance observers

The earliest applications of disturbance observers appeared in the line of re-search by Ohnishi and colleagues [33,35] and Lee and Tomizuka [39] for mechatronics motion control systems. The first sensorless reaction torque estimation for manipulators with multiple degrees of freedom in industrial settings was proposed by Murakami anf colleagues [40]. The first extension of linear disturbance observers to robotic manipulators with nonlinear dynamics appeared in the seminal work by Chen and coworkers [41], where a stability proof for disturbance observer tracking error convergence for robotic manipulators with revolute joints and planar kinematic configurations was presented. More recently, Chen and coworker's nonlinear disturbance observer formulation was generalized to robotic manipulators with non-planar kinematic structure and with both revolute and prismatic joints in the work by Mohammadi and colleagues [42].

In this section, the underlying architecture and properties of linear and non-linear disturbance observers are briefly reviewed. The first linear disturbance observer was proposed in [33,35] for robust motion control systems and has since motivated a long line of research in both controls and robotics communities. Fig. 5.1 depicts the block diagram of a disturbance observer-based motion control system. The torque τ represents the control input. Moreover, the parameters J and L_d represent mass/inertia and gain of the disturbance observer, which is a design parameter, respectively. The unknown disturbance τ_d lumps the effect of disturbances such as harmonics and friction; namely,

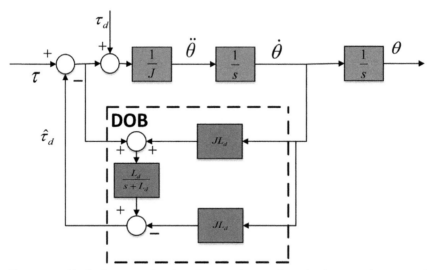

Figure 5.1 Block diagram of a disturbance observer-based robust motion control system proposed in Refs. [33,35]. The effect of external disturbances and plant uncertainties, such as variations in the inertia parameters, are lumped by the unknown disturbance τ_d. The control input is represented by the torque τ. The parameters J and L_d are the nominal inertia and gain of the disturbance observer, respectively.

$$\tau_d = \tau_{d,f} + \tau_{d,h},$$

where $\tau_{d,f}$ and $\tau_{d,h}$ are the friction and harmonic disturbances, respectively. Similarly, other disturbances can be incorporated in the lumped disturbance τ_d. The role of the disturbance observer is to estimate the lumped disturbance torque τ_d from the input τ and the state $\dot{\theta}$. The dynamics of the motion control system are governed by

$$J\ddot{\theta} = \tau + \tau_d - \hat{\tau}_d$$

Feeding forward the disturbance estimate $\hat{\tau}_d$ suppresses the adverse effect of τ_d on the motion control system in Fig. 5.1. In the ideal case, when $\hat{\tau}_d = \tau_d$, the lumped disturbance will be canceled out and the control input τ acts on the nominal control system with no disturbances. As it can be seen from the block diagram in Fig. 5.1,

$$\hat{\tau}_d = \frac{L_d}{s + L_d} JL_d\dot{\theta} + \tau - \hat{\tau}_d JL_d\dot{\theta} = \frac{L_d}{s + L_d}\left(Js\dot{\theta} + \tau - \hat{\tau}_d\right)$$

Therefore, the disturbance estimate satisfies

$$\hat{\tau}_d = \frac{L_d}{s + L_d}\tau_d = \frac{1}{\dfrac{1}{L_d S + 1}}\tau_d$$

If the lumped disturbance term τ_d remains within the bandwidth of the disturbance observer, it can be accurately estimated by the output of the disturbance observer, namely, $\hat{\tau}_d$. Indeed, larger disturbance observer gains give rise to smaller tracking errors during the steady state.

Given the nonlinear dynamics of a robot and nominal control law $u_n x$, the underlying mechanism of nonlinear disturbance observers operation can be explained as follows. Using the state and input information, the nonlinear disturbance observer estimates the disturbance $d(t)$, which is degrading the performance of the nominal controller. The estimated disturbance $\hat{d}$ is then added to the nominal control input in the following feedforward manner:

$$u = u_n(x) - \hat{d}. \tag{5.1}$$

In the ideal case, when $\hat{d} = dt$, the disturbances acting on the system would be canceled out and the nominal control input $u_n x$ would achieve the desired objectives, without the need for modifying the nominal controller. The nonlinear disturbance observer dynamics for a robot with inertia matrix M q, centrifugal/Coriolis matrix C q, q, gravity vector G q, and control input u are given by (see [28,41,43] for the details of its derivation)

$$\dot{z} = -L_d^\varepsilon(q)z + L_d^\varepsilon(\dot{q})C(q,\dot{q})\dot{q} + G(q) - u - p^\varepsilon(\dot{q})\Sigma, \dot{d} = z + p^\varepsilon(\dot{q}), \quad (5.2)$$

where z R^N, which is an auxiliary vector for removing the need for acceleration measurements, is the *state* of the nonlinear disturbance observer. The *auxiliary vector* and the *gain matrix* of the nonlinear disturbance observer are given by

$$p^\varepsilon(\dot{q}) = X_\varepsilon^{-1}\dot{q}, \; L_d^\varepsilon(q) = X_\varepsilon^{-1}M^{-1}(q). \quad (5.3)$$

respectively, where X_ε $R^{N\times N}$ is a constant symmetric and positive definite matrix depending on a positive constant ε. For simplicity of exposition, we let $X_\varepsilon = \varepsilon I_N$. Finally, NDOB disturbance tracking error is defined to be

$$e_d := \hat{d} - d(t). \quad (5.4)$$

When $\hat{d} = dt$, namely, in the ideal case, $e_d = 0$.

As it has been shown in Refs. [29,44], disturbance observers for fully actuated mechanical systems enjoy several properties including ultimate boundedness of disturbance tracking error in the presence of disturbances with a bounded rate of change, bounded-input bounded-output (BIBO) stability, and semi/quasi-passivity properties.

3 Disturbance observers in upper extremity robotic rehabilitation

The human arm or the upper limb, which extends from the shoulder to the fingertips, consists of the upper arm, the forearm, and hand, which are connected by shoulder, the elbow, and the wrist, respectively. In upper extremity robotic rehabilitation, the mechanical structures of the robots can be categorized into end-effector-based robots and exoskeleton-type robots [45]. Though being cost-effective, end-effector-based robots such as MIT-MANUS [46], which get attached to the patient's forearm or hand at a single point, cannot achieve anthropomorphic trajectory tracking a the human joint level. On the other hand, exoskeleton-type upper extremity rehabilitation robots such as HEXORR [47] and ARMin [13], which are either wall-grounded or mobile, get attached to the patient's body along the targeted limb. These robots are capable of inducing exact trajectory tracking at the human joint level, provided that they are properly aligned with the patient's joints.

In upper extremity rehabilitation robots, different training modalities including passive, active-resisted, and active-assisted modes [23,45]. During the passive training mode, the patient's arm is moved through a predetermined path by the robotic device. In the active-resisted training mode, the robotic device resists the patient's movements, whereas in the active-assisted training mode, the robot will provide partial assistance during the movement if the patient is not capable of moving his/her upper extremity limbs (An LMI-based formulation for determining the optimal X_ε has been proposed in [42]; The semi/quasi-passivity property of nonlinear disturbance observers was first reported by Mohammadi *et al.* in [29]).

In this section, the applications of disturbance observers to different types of upper extremity rehabilitation devices, which either target the upper arm/forearm or the hand regions, are reviewed.

3.1 Upper arm/forearm rehabilitation

Disturbance observers can be used across different levels of hierarchy in the upper arm/forearm rehabilitation robot control systems. In the low-level of rehabiliation robot control hierarchies [37,38], where the control objective is to track the desired joint trajectories, disturbance observers enable the actuators to accurately track the desired trajectories by suppressing the deteriorating effect of disturbances.

Park and coworkers [48] present a robot-assisted bimanual shoulder flexion rehabilitation system. A combination of surface electromyography and bimanual mirror imaging motion was used for induction of voluntary stimulation to hemiplegic patients' neuro-musculoskeletal system. In order to have a back-drivable operation for assistive rehabilitation, Park and coworkers employed linear disturbance observers for compensating the impedance of the actuator. Having a closed-loop dynamics with low mechanical impedance removes the need for generating additional torques in order to overcome the actuator resistive torque. In [49], Q. Wu and H. Wu propose a multi-modal control strategy, including both patient-passive and patient-cooperative training modes, for a therapeutic upper limb exoskeleton with seven actuated degrees of freedom. In order to guarantee position control precision during the passive training mode, the authors used a disturbance observer along with an adaptive sliding mode controller. Chen and collaborators use a combination of fuzzy approximation algorithms and nonlinear disturbance observers for robust control and sensorless torque estimation in their upper-body exoskeleton [50,51]. Chen and coworkers verify the effectiveness of their proposed control methodology on an upper limb exoskeleton system consisting of an arm with a total of five degrees of freedom. The five single-axis revolute joints in the robotic arm exoskeleton can mimic the motion of a human arm with wrist radial ulnar motion, shoulder and elbow extension-flexion, shoulder adduction-abduction, and forearm supination-pronation. One of the features in fuzzy-based nonlinear disturbance observer of Chen and coworkers is its capability of suppressing the deteriorating effect of deadzone and actuator saturation in the upper-body exoskeleton.

In both the mid-level and high-level of rehabilitation robot control hierarchies [37,38], where the objective is to recognize the intent of the patient and map the patient's intents to proper device outputs, there is a need for estimating the human–robot interaction forces/torques. One of the main applications of disturbance observers in mechatronics is in the area of sensorless force/torque estimation. Consequently, disturbance observers can be efficiently employed in the mid and high-level of hierarchies in the rehabilitation robot control system. In the line of work by Ugurlu and coworkers [52,53], a framework for sensorless torque estimation in wearable exoskeletons for power augmentation and rehabilitation is introduced. The proposed framework relies on accurate identification and compensation of the joint-level disturbance torques caused by various types of joint frictions and the joint gravitational loads. After compensating the gravitational and frictional loads at each joint, Ugurlu and coworkers employ linear disturbance observers to estimate the external torque exerted at each joint. The estimation of human-wearer torques, which were estimated by the linear disturbance observers, closely tracks the measured torques by EMG sensors. This framework removes the need for having expensive torque sensing units in rehabilitation and assistive robots. Along the same line of removing the need for having built-in torque sensing modules, Kim and coworkers [27,54] employ a passivity-based framework for design of nonlinear disturbance observers in powered upper-limb robotic exoskeletons in industrial settings. The proposed passivity-based disturbance observer possesses several features such as allowing for the nonlinearities in the robotic exoskeleton dynamics, simultaneous embedding of human operator and environmental interactions in the closed-loop control, and providing the user with nominal model selection capability while wearing the exoskeleton.

3.2 Hand rehabilitation

Various research papers have been published in the field of hand exoskeletons and haptic gloves with applications in rehabilitation [55]. However, compared to the other well-developed force feedback rehabilitation devices for the other upper and lower extremities, developing hand rehabilitation exoskeletons along with their control systems still face various obstacles. Indeed, the requirements for tactile sensing, performing complex motion patterns, and smaller size are among the most important challenges for having efficient robotic hand rehabilitation devices [55]. One of the obstacles of having robotic devices that are similar to the human fingers is that the human finger joint structures are complicated and difficult to completely mimic in mechanical design. For instance, the anatomical structure of index finger consists of three phalanges, namely, distal, middle, and proximal phalanges. Therefore, there is a need for lightweight finger robotic wearable devices that mimic the human finger's basic motion according to the natural sequence of finger muscle activation pattern. Furthermore, since force control for grasping can be achieved using tactile feedback, the limited space in hand exoskeletons makes the introduction of a tactile sensor on the robotic finger surface area very challenging. The need for tracking precision and sensor torque/force estimation justify using disturbance observers in robotic hand exoskeletons.

In the line of work by Lemerle and coworksers [56,57], the control strategy for robotic finger rehabilitation is based on the anatomical observation that for tendon therapy, avoiding gap formation and break of the suture is essential. Based on this assessment, Lemerle and coworkers envision a bilateral control structure, where the master system is controlled by the patients themselves. To implement this control strategy, a four-channel bilateral architecture with linear disturbance observers, which is due to Iida and Ohnishi [58,59], is employed. Jo and Bae [60] employ disturbance observers in the low-level of control hierarchy in their force-controllable hand exoskeleton with compact actuator modules. In order to achieve precise tracking by their series elastic actuators, they employ disturbance observers to cancel the deteriorating effect of disturbances acting on the hand exoskeleton control system.

In the line of work by Popescu and coworkers [61–63], a robotic hand exoskeleton with an anthropomorphic kinematic structure with five articulated fingers has been built, where each finger consists of three phalanges in order to have an anthropomorphic contact with the patient's hand. In order to remove the complications associated with having a distributed sensor network for the finger exoskeleton, Popescu and coworkers [61] employ a hierarchical velocity-force-disturbance estimation scheme, where a cascade closed-loop control strategy is implemented with velocity and force observers. In order to achieve actuator compliance compensation, a disturbance observer is employed along with the velocity and force observers. Using this hierarchical torque estimator, the intentions of the human can also be indicated.

In the robotics rehabilitation literature, several robotic devices, among which MIT-Manus [46] is one of the most prominent, target wrist rehabilitation trainings. In [64–67], Takaiwa and collaborators introduce a pneumatic parallel manipulator for patients' wrist rehabilitation. Having pneumatic actuators, their pneumatic wrist rehabilitation device can perform anthropomorphic human wrist joint motions due to the air compressibility and parallel manipulator's feature of multiple degrees of freedom. Impedance control system is introduced to realize several rehabilitation modes. In the low-level of their rehabilitation robot control hierarchy, Takaiwa and collaborators use linear disturbance observers to suppress the unwanted disturbances due to the air nonlinear pressure.

In [68,69], Saadatzi and collaborators use the general nonlinear disturbance observer design framework in Mohammadi and coworkers [42] for torque estimation in their wrist rehabilitation robotic device, which has a non-planar kinematic configuration. In their experiments, Saadatzi and collaborators noticed that when nonlinear disturbance observers are used with accurate model parameters, better average tracking errors in comparison with inverse dynamics-based and other simpler control approaches are achieved. In the line of work by Pehlivan and coworkers [70,71], nonlinear disturbance observers are employed for estimating the patient input torques to RiceWrist-S robotic wrist rehabilitation device.

Upper extremity wire-driven rehabilitation robots have lower moving inertia and a lower construction cost. Furthermore, unlike end-effector-based rehabilitation robots do not suffer from having poor workspace. Niu and coworkers [72,73] have developed a wire-driven upper extremity rehabilitation robot that can operate in a three-dimensional space. In order to deliver task-oriented rehabilitation exercises for the patient's upper-limb, they design a sliding mode control scheme along with a nonlinear disturbance observer to suppress the effect of disturbances. The role of the nonlinear disturbance observer is to reduce the chattering, which is induced due to using sliding mode controllers, and its degrading effect on the robot actuators (This four-channel bilateral control architecture was later extended to nonlinear settings by Mohammadi et al. [59]).

4 Disturbance observers in lower extremity robotic rehabilitation

Lower limb robotic rehabilitation devices target the knee, hip, and/or ankle joint levels. Equipped with proper actuators and the capability of sensing the orientation of the patient's joints, ground reaction forces, position of the limbs, and the patient's muscular activity levels, the control system of the lower limb rehabilitation robots can drive and/or induce the desired movement of the patient's limbs.

Lower limb rehabilitation robots can be classified into two main groups [74]; namely, immobile robots such as Lokomat [21] and LOPES [15], and over- ground rehabilitation robots. Immobile rehabilitation robots, which include body-weight-support footplate trainers, treadmill-based body-weight-support systems, and treadmill-based exoskeletons, provide proper gait training to the patients in a confined and fixed area. In the second group, whose design philosophy is based on increasing the independence of gait exercises, rehabilitation robots allow the patients to walk over the ground independently. A recent study in [75] has demonstrated that using untethered knee joint orthoses can enhance the speed of post-stroke patients' gaits.

In this section, the applications of disturbance observers to different types of lower extremity rehabilitation devices, which either target the ankle, the knee, or both regions, are reviewed.

4.1 Ankle rehabilitation

The human ankle often undergoes large loads that might reach up to several times of the body weight. Consequently, there is a high possibility of ankle injuries such as severe ankle sprains, which is among the most common musculoskeletal injuries requiring intensive rehabilitation exercises [76]. In addition to severe ankle sprains, there are certain other ankle conditions such as "foot drop" that arise due to neuromuscular diseases. The foot drop condition refers to the situation where the patients fail to rotate their foot upwards during walking due to the loss of control for ankle dorsiflexor muscles. In order to permit higher mobility, the actuators that are employed in wearable ankle robots are often of lower inertia. Furthermore, the ankle robot actuators are required to be inherently backdrivable to ensure the patient's safety of the user.

In the work by Yu and coworkers [77], a safe and stable human–robot interaction control approach for an ankle rehabilitation robot, which is actuated by a series elastic actuator, is presented. In this control approach, the proposed controller achieves human motion compensation using a disturbance observer, where there is no need of feeding back the acceleration of the assistive robot in the closed-loop control system. In the work by Kotina and coworkers [78], human postural sway is controlled by employing an active disturbance rejection controller. Although obtaining a realistic model of the musculoskeletal activity of the ankle joint during human postural sway is a very challenging task, the disturbance estimation method in this work achieves excellent performance for the nonlinear, uncertain, and time-varying dynamic systems in simulations.

In the work by Tsoi and coworkers [79,80], a force-based variable impedance controller for an ankle rehabilitation parallel robot with multiple degrees of freedom has been developed. The proposed impedance controller, which employs disturbance observers in its internal structure for decoupling of the forces, only employs joint position information in order to regulate the total vertical force applied by the actuators. In the work by Jamwal and coworkers [81], a lightweight parallel robot with pneumatic muscle actuators, which are placed in parallel to the shin bone, is developed for ankle rehabilitation exercises. In order to be able to control the ankle rehabilitation robotic device, the authors employ a fuzzy logic-based control approach along with a disturbance observer to counter the unknown dynamics of the pneumatic muscle actuators in their design. Along the same line of research on developing ankle rehabilitation robotic devices, Ai and coworkers [82] developed a parallel ankle rehabilitation robot with pneumatic muscles. Pneumatic muscles, while being backdrivable and flexible, demonstrate highly nonlinear characteristics during patient's walking. Furthermore, the human–robot interaction forces, if unknown, act as disturbances and uncertainties in the rehabilitation robot control system. Ai and coworkers [82] use an adaptive backstepping sliding mode control scheme along with a disturbance observer to suppress the deteriorating effect of pneumatic muscle actuator nonlinearities and human–robot interaction forces.

4.2 Knee and knee-ankle rehabilitation

One of the vital movements in the knee rehabilitation exercises is that of the flexion-extention, where the knee bends during the former and extends during the latter. Knee rehabilitation robots, compared to conventional non-robotic physiotherapy procedures, enjoy repeatable, continuous, and smooth movements of the patient's knee joint in clinical settings. Similar to upper extremity rehabilitation, knee rehabilitation robotic devices can be categorized into two main groups; namely, end-effector-based robotic manipulators [7], and knee powered exoskeletons/orthoses [83]. The amount of power that should be provided to ensure a given task is determined by means of an appropriate control law. In powered knee orthoses, the exercise is determined by a desired therapeutic trajectory, provided by a physiotherapist or according to the intention of the patient to improve his/her performance. On-board sensors or detection of the nueromuscular activity levels can be used to estimate the intention of the patient.

In the line of work by Amirat, Mohammed, and collaborators [84–86], a model which captures the integrated dynamics of human lower-limb and orthosis is developed. In order to improve the control performance in terms of human muscular effort, a nonlinear disturbance observer for estimating the patient's muscular torque has been developed. The role of the nonlinear disturbance observer is to suppress the effect of dynamical uncertainties in the patient's muscular torque. In Ref. [85], two types of sliding mode controllers, namely, of basic and robust terminal types, have been employed along with the nonlinear disturbance observer in the lower limb powered orthosis control system. In both of these cases, it has been verified that nonlinear disturbance observer improves the tracking precision as well as robustness to external disturbances.

In the work by Khamar and Edrisi [87], a control scheme for a knee end effector-based rehabilitation robot with one revolute joint, which moves in the sagittal plane, is proposed. The constructed knee rehabilitation robot possesses a linear elastic actuator that is connected to both the lower and the upper parts of the knee brace, allowing for having a low–impedance interface for the patient. In order to estimate the patient's muscular torque, increasing the tracking precision, and extending the bandwidth of the controller, a robust backstepping sliding control along with a nonlinear disturbance observer is employed in the control structure in Ref. [87].

Recently, some researchers have started using cable-driven rotary series elastic actuators [88,89] in their rehabilitation robotic devices. In order to attenuate the effect of variable friction in Bowden cables that are used in these robotic mechanisms, Lu and coworkers [88] employ disturbance observers to estimate the unknown friction torques and cancel their effect via feedforward compensation.

5 Disturbance observers in robotic telerehabilitation

Physical telerehabilitation research in both academia and industry has been on the rise over the last 2 decades [90,91]. Without decreasing the quality of the patient's treatment, the desired objective of telerehabilitation is to provide the patient with rehabilitation therapy in home, and to assess the patient's rehabilitation progress through the Internet. Indeed, due to the costly nature of intensive patient rehabilitation, robotic telerehabilitation systems can effectively provide the patient with consistent and repeatable rehabilitation exercise routines. To perform online rehabilitation exercises with live assistance or cooperation of a physiotherapist, another robot for capturing the therapist's inputs is required in the telerehabilitation tasks.

There are two main paradigms for telerehabilitation. In the first paradigm, the patient performs independent rehabilitation exercises through virtual reality (VR) environments by means of using wearable robotic devices such as hand exoskeletons and sensorized gloves, which capture the patient's motion kinematics and provide corrective feedback [92,93]. In VR-based telerehabilitation systems that rely on reflecting corrective force feedback to the patient, the aim is to feedback the force from the virtual environment to the patient in order to guide the patient limbs on the proper training trajectories. In this telerehabilitation paradigm, there is a minor involvement by the clinician, who assesses the patient's progress and tunes the parameters of the VR-based rehabilitation exercises from the remote site. In other words, while the physical guidance is provided by the robot, the clinician is responsible for adjusting the high-level exercise parameters and providing verbal feedback to the patient.

In the second rehabilitation paradigm, the patient needs strong supervision from the clinician to perform rehabilitation exercises. Master-slave bilateral teleoperation setups belong to this paradigm [94–96], where the therapist and the patient can interact with each other for various functional and movement therapies. In this paradigm, the physiotherapist can guide the targeted limb along a desired motion trajectory, where the exercise parameters can be adjusted based on the patient's feedback in real time. In bilateral telerehabilitation systems, haptic feedback can be provided in either passive or active modes [95,97]. In the passive mode, the patient's limbs are moved by the haptic device, which is directly controlled by the therapist. On the other hand, in the active mode, the patient actively moves his/her hand while the therapist, depending on the patient's state and the type of the rehabilitation training, assists or resists the movements remotely. For experienced therapists, haptic feeling, provided that it is transparent, will be more beneficial than quantitative measures for training the patient limbs remotely. In this work, we only discuss the applications of disturbance observers to the second telerehabiliation paradigm.

In bilateral telerehabilitation setups, there are inherent problems such as random data packet losses and variable time delays [99,100] that hinder the successful application of the assistive rehabilitation systems to telerehabilitation setups. In assistive rehabilitative systems, which are required to have a very high sampling frequency rate, the issue of random packet loss can lead to bilateral system instability. Furthermore, the limited computation power of commercial micro-processors might exacerbate the real-time control of bilateral telerehabilitation system further.

In the line of work by Bae and coworkers [101–103], a disturbance observer-based control approach, which is inspired from the network-based control literature, is used to suppress the adverse effect of random packet losses in bilateral telerehabilitation systems. Bae and collaborators model the packet loss of the control command, which is issued from the local host controller, to the rehabilitation robot actuator as an external disturbance. Using disturbance observers, they estimate the approximate packet loss values and reconstruct the lost commands from the local actuator in a feedforward manner (It is remarked that one of the first applications of nonlinear disturbance observers for estimating haptic interaction forces appeared in the work by Gupta and O'Malley [98], where a nonlinear disturbance observer was used in a single-degree-of-freedom haptic interface. Gupta and O'Malley demonstrated that using disturbance observers in haptic interfaces increase the haptic interaction fidelity by improving the transparency bandwidth of the interface).

Another inherent issue in bilateral telerehabilitation robotic systems is due to the presence of variable time delays in the communication channel. Natori and coworkers [104,105] have developed a novel type of disturbance observers, called the communication disturbance observers that estimate the time delay in the communication channel and compensate for the communication time delays. Communication disturbance observers have been shown to be more effective than the Smith-predictor-based delay compensation schemes since this type of disturbance observers works without having to rely on communication time-delay model. Furthermore, it has been demonstrated that communication disturbance observers are also capable of handling time-varying delays. Recently, Zhang and Tomizuka [106] have developed a network-based gait rehabilitation system for improving mobility and telerehabilitation. On the patient's side, assistive torques are provided by a rotary elastic actuator of compact size, which is controlled over a local wireless network. However, the presence of variable time delays in both controller-actuator and sensor-controller channels causes performance degradation and destabilization in more severe cases. In the work by Zhang and Tomizuka [106], the communication time delay is compensated by employing a discrete-time communication disturbance observer.

6 Conclusion and further remarks

Having seamless interaction between rehabilitation robots and patients as well as achieving effective rehabilitation training, developing efficient and safe control algorithms is of vital importance for restoration of motion in both upper and lower extremities. A prevalent challenge in control of rehabilitation robotic devices is the simultaneous presence of disturbances and persistent human inputs during human–robot interaction. Moreover, in any application where the human and the robotic device are interacting with each other, measuring the interaction forces becomes important for guaranteeing the performance and safety of the rehabilitation robot control system. However, there are inherent problems when force sensors are employed in the rehabilitation robot control system. A recent cost-effective alternative to inclusion of force sensors in the rehabilitation robotic device is based on sensorless force estimation, whereby the forces due to the human–robot interaction are estimated using disturbance observers. As an alternative for force/torque sensors, disturbance observers can provide the robot control system with information about the patient–robot interaction forces as well as unknown disturbances. This chapter presented an overview of disturbance observer applications in rehabilitation robotics for upper/lower extremity rehabilitation and telerehabilitation.

References

[1] N. Hogan, H.I. Krebs, Physically interactive robotic technology for neuro-motor rehabilitation, Prog. Brain Res. 192 (2011) 59–68.
[2] N. Norouzi-Gheidari, P.S. Archambault, J. Fung, Effects of robot-assisted therapy on stroke rehabilitation in upper limbs: systematic review and meta- analysis of the literature, J. Rehabil. Res. Dev. 49 (4) (2012).
[3] H. Krebs, B.Volpe, Rehabilitation robotics, Handbook of Clinical Neurology, vol. 110, 2013, pp. 283–294.
[4] G. Kwakkel, B.J. Kollen, H.I. Krebs, Effects of robot-assisted therapy on upper limb recovery after stroke: a systematic review, Neurorehabil. Neural Repair 22 (2) (2008) 111–121.
[5] S.Viteckova, P. Kutilek, M. Jirina, Wearable lower limb robotics: a review, Biocybern. Biomed. Eng. 33 (2) (2013) 96–105.
[6] S.A. Kolakowsky-Hayner, J. Crew, S. Moran, A. Shah, Safety and feasibility of using the ekso TM bionic exoskeleton to aid ambulation after spinal cord injury, J. Spine 4 (003) (2013).

[7] E. Akdoğan, E. Taçgın, M.A. Adli, Knee rehabilitation using an intelligent robotic system, J. Intell. Manuf. 20 (2) (2009) 195.

[8] Y.-L. Park, B.-r. Chen, N.O. Pérez-Arancibia, D. Young, L. Stirling, R.J. Wood, E.C. Goldfield, R. Nagpal, Design and control of a bio-inspired soft wearable robotic device for ankle–foot rehabilitation, Bioinspir. Biomim. 9 (1) (2014) 016007.

[9] H. Kawasaki, S. Ito, Y. Ishigure, Y. Nishimoto, T. Aoki, T. Mouri, H. Sakaeda, M. Abe, Development of a hand motion assist robot for rehabilitation therapy by patient self-motion control, in: 2007 IEEE 10th International Conference on Rehabilitation Robotics, 2007, pp. 234–240.

[10] S. Balasubramanian, J. Klein, E. Burdet, Robot-assisted rehabilitation of hand function, Curr. Opin. Neurol. 23 (6) (2010) 661–670.

[11] S. Jezernik, G. Colombo, T. Keller, H. Frueh, M. Morari, Robotic orthosis lokomat: a rehabilitation and research tool, Neuromodulation 6 (2) (2003) 108–115.

[12] H.I. Krebs, J.J. Palazzolo, L. Dipietro, M. Ferraro, J. Krol, K. Rannekleiv, B.T. Volpe, N. Hogan, Rehabilitation robotics: performance-based progressive robot-assisted therapy, Autonomous Robots 15 (1) (2003) 7–20.

[13] T. Nef, R. Riener, ARMin-design of a novel arm rehabilitation robot, in: Ninth International Conference on Rehabilitation Robotics, 2005. ICORR 2005, 2005, pp. 57–60.

[14] A. Gupta, M.K. O'Malley, Design of a haptic arm exoskeleton for training and rehabilitation, IEEE/ASME Trans. Mechatron. 11 (3) (2006) 280–289.

[15] J.F. Veneman, R. Kruidhof, E.E. Hekman, R. Ekkelenkamp, E.H. Van Assel- donk, H. Van Der Kooij, Design and evaluation of the LOPES exoskeleton robot for interactive gait rehabilitation, IEEE Trans. Neural Syst. Rehabil. Eng. 15 (3) (2007) 379–386.

[16] A. Pennycott, D. Wyss, H. Vallery, V. Klamroth-Marganska, R. Riener, Towards more effective robotic gait training for stroke rehabilitation: a review, J. Neuroeng. Rehabil. 9 (1) (2012) 65.

[17] H. Vallery, A. Duschau-Wicke, R. Riener, Generalized elasticities improve patient-cooperative control of rehabilitation robots, in: 2009 IEEE International Conference on Rehabilitation Robotics, 2009, pp. 535–541.

[18] M. Mihelj, T. Nef, R. Riener, A novel paradigm for patient-cooperative control of upper-limb rehabilitation robots, Adv. Robotics 21 (8) (2007) 843–867.

[19] E.T. Wolbrecht, D.J. Reinkensmeyer, J.E. Bobrow, Pneumatic control of robots for rehabilitation, Int. J. Robotics Res. 29 (1) (2010) 23–38.

[20] S. Jezernik, G. Colombo, M. Morari, Automatic gait-pattern adaptation algorithms for rehabilitation with a 4-dof robotic orthosis, IEEE Trans. Robotics Automation 20 (3) (2004) 574–582.

[21] L. Lunenburger, G. Colombo, R. Riener, V. Dietz, Biofeedback in gait training with the robotic orthosis Lokomat, in: The 26th Annual International Conference of the IEEE Engineering in Medicine and Biology Society, vol. 2, 2004, pp. 4888–4891.

[22] G. Kurillo, M. Gregorič, N. Goljar, T. Bajd, Grip force tracking system for assessment and rehabilitation of hand function, Technol. Health Care 13 (3) (2005) 137–149.

[23] R. Colombo, F. Pisano, S. Micera, A. Mazzone, C. Delconte, M.C. Carrozza, P. Dario, G. Minuco, Robotic techniques for upper limb evaluation and rehabilitation of stroke patients, IEEE Trans. Neural Syst. Rehabil. Eng. 13 (3) (2005) 311–324.

[24] C. Mitsantisuk, K. Ohishi, S. Katsura, Estimation of action/reaction forces for the bilateral control using Kalman filter, IEEE Trans. Ind. Electron. 59 (11) (2011) 4383–4393.

[25] D.C. Rucker, R.J. Webster, Deflection-based force sensing for continuum robots: a probabilistic approach, in: 2011 IEEE/RSJ International Conference on Intelligent Robots and Systems, 2011, pp. 3764–3769.

[26] A. Wahrburg, E. Morara, G. Cesari, B. Matthias, H. Ding, Cartesian contact force estimation for robotic manipulators using kalman filters and the generalized momentum, in: 2015 IEEE International Conference on Automation Science and Engineering (CASE), 2015, pp. 1230–1235.

[27] M.J. Kim, Y.J. Park, W.K. Chung, Design of a momentum-based disturbance observer for rigid and flexible joint robots, Intell. Serv. Robotics 8 (1) (2015) 57–65.

[28] W.-H. Chen, Disturbance observer based control for nonlinear systems, IEEE/ASME Trans. Mechatron. 9 (4) (2004) 706–710.

[29] A. Mohammadi, H.J. Marquez, M. Tavakoli, Nonlinear disturbance observers: design and applications to Euler-Lagrange systems, IEEE Control Syst. Mag. 37 (4) (2017) 50–72.

[30] W.-H. Chen, J. Yang, L. Guo, S. Li, Disturbance-observer-based control and related methods—an overview, IEEE Trans. Ind. Electron. 63 (2) (2016) 1083–1095.

[31] S. Li, J. Yang, W.-H. Chen, X. Chen, Disturbance Observer-Based Control: Methods and Applications, Boca Raton, FL, CRC Press, (2014).

[32] E. Sariyildiz, K. Ohnishi, Stability and robustness of disturbance-observer-based motion control systems, IEEE Trans. Ind. Electron. 62 (1) (2015) 414–422.

[33] K. Ohnishi, M. Shibata, T. Murakami, Motion control for advanced mechatronics, IEEE/ASME Trans. Mechatron. 1 (1) (1996) 56–67.

[34] A. Radke, Z. Gao, A survey of state and disturbance observers for practitioners, in: Proc. 2006 Amer. Contr. Conf., 2006, pp. 5183–5188.

[35] K. Ohnishi, A new servo method in mechatronics, Trans. Japan. Soc. Elect. Eng. 107-D (1987) 83–86.

[36] K. Ohishi, Torque-speed regulation of dc motor based on load torque estimation, in: IEEJ International Power Electronics Conference, IPEC-TOKYO, 1983-3, vol. 2, 1983, pp. 1209–1216.

[37] H.A. Varol, F. Sup, M. Goldfarb, Multiclass real-time intent recognition of a powered lower limb prosthesis, IEEE Trans. Biomed. Eng. 57 (3) (2009) 542–551.

[38] M.R. Tucker, J. Olivier, A. Pagel, H. Bleuler, M. Bouri, O. Lambercy, J. del, R. Millán, R. Riener, H. Vallery, R. Gassert, Control strategies for active lower extremity prosthetics and orthotics: a review, J. Neuroeng. Rehabil. 12 (1) (2015) 1.

[39] H.S. Lee, M. Tomizuka, Robust motion controller design for high-accuracy positioning systems, IEEE Trans. Ind. Electron. 43 (1) (1996) 48–55.

[40] T. Murakami, F. Yu, K. Ohnishi, Torque sensorless control in multidegree- of-freedom manipulator, IEEE Trans. Ind. Electron. 40 (2) (1993) 259–265.

[41] W.-H. Chen, D.J. Ballance, P.J. Gawthrop, J. O'Reilly, A nonlinear disturbance observer for robotic manipulators, IEEE Trans. Ind. Electron. 47 (4) (2000) 932–938.

[42] A. Mohammadi, M. Tavakoli, H. Marquez, F. Hashemzadeh, Nonlinear disturbance observer design for robotic manipulators, Contr. Eng. Prac. 21 (3) (2013) 253–267.

[43] A. Mohammadi, H.J. Marquez, M. Tavakoli, Nonlinear disturbance observers: design and applications to Euler-Lagrange systems, IEEE Contr. Syst. 37 (4) (2017) 50–72.

[44] W.-H. Chen, L. Guo, Analysis of disturbance observer based control for nonlinear systems under disturbances with bounded variation, in: Proceedings of International Conference on Control, 2004, pp. 1–5.

[45] M. Babaiasl, S.H. Mahdioun, P. Jaryani, M. Yazdani, A review of technological and clinical aspects of robot-aided rehabilitation of upper-extremity after stroke, Disabil. Rehabil. Assist. Technol. 11 (4) (2016) 263–280.

[46] N. Hogan, H.I. Krebs, J. Charnnarong, P. Srikrishna, A. Sharon, Mit-manus: a workstation for manual therapy and training, in: [1992] Proceedings IEEE International Workshop on Robot and Human Communication. IEEE, 1992, pp. 161–165.

[47] C.N. Schabowsky, S.B. Godfrey, R.J. Holley, P.S. Lum, Development and pilot testing of HEXORR: hand exoskeleton rehabilitation robot, J. Neuroeng. Rehabil. 7 (1) (2010) 36.

[48] K. Park, S. Kwon, J. Kim, B. Rim, Bimanual shoulder flexion system with surface electromyography for hemiplegic patients after stroke: a preliminary study, in: 2011 IEEE International Conference on Rehabilitation Robotics. IEEE, 2011, pp. 1–6.

[49] Q. Wu, H. Wu, Development, dynamic modeling, and multi-modal control of a therapeutic exoskeleton for upper limb rehabilitation training, Sensors 18 (11) (2018) 3611.

[50] Z. Chen, Z. Li, C.P. Chen, Disturbance observer-based fuzzy control of uncertain mimo mechanical systems with input nonlinearities and its application to robotic exoskeleton, IEEE Trans. Cybern. 47 (4) (2017) 984–994.

[51] Z. Li, C.-Y. Su, L. Wang, Z. Chen, T. Chai, Nonlinear disturbance observer-based control design for a robotic exoskeleton incorporating fuzzy approximation, IEEE Trans. Ind. Electron. 62 (9) (2015) 5763–5775.

[52] B. Ugurlu, M. Nishimura, K. Hyodo, M. Kawanishi, T. Narikiyo, A framework for sensorless torque estimation and control in wearable exoskeletons, in: 2012 12th IEEE International Workshop on Advanced Motion Control (AMC). IEEE, 2012, pp. 1–7.

[53] B. Ugurlu, M. Nishimura, K. Hyodo, M. Kawanishi, T. Narikiyo, Proof of concept for robot-aided upper limb rehabilitation using disturbance observers, IEEE Trans. Human Machine Syst. 45(1) (2015) 110–118.

[54] M.J. Kim, W. Lee, J.Y. Choi, Y.S. Park, S.H. Park, G. Chung, K.-L. Han, I.S. Choi, I.H. Suh, Y. Choi et al., Powered upper-limb control using passivity-based nonlinear disturbance observer for unknown payload carrying applications, in: 2016 IEEE International Conference on Robotics and Automation (ICRA). IEEE, 2016, pp. 2340–2346.

[55] P. Heo, G.M. Gu, S.-j. Lee, K. Rhee, J. Kim, Current hand exoskeleton technologies for rehabilitation and assistive engineering, Int. J. Precis. Eng. Manuf. 13 (5) (2012) 807–824.

[56] S. Lemerle, S. Fukushima, Y. Saito, T. Nozaki, K. Ohnishi, Wearable finger exoskeleton using flexible actuator for rehabilitation, in: 2017 IEEE International Conference on Mechatronics (ICM), 2017, pp. 244–249.

[57] S. Lemerle, T. Nozaki, K. Ohnishi, Design and evaluation of a remote actuated finger exoskeleton using motion-copying system for tendon rehabilitation, IEEE Trans. Ind. Inform. 14 (11) (2018) 5167–5177.

[58] W. Iida, K. Ohnishi, Reproducibility and operationality in bilateral teleoperation, in: The 8th IEEE International Workshop on Advanced Motion Control, 2004. AMC'04, 2004, pp. 217–222.

[59] A. Mohammadi, M. Tavakoli, H. Marquez, Disturbance observer-based control of non-linear haptic teleoperation systems, IET Contr. Th. Applicat. 5 (18) (2011) 2063–2074.

[60] I. Jo, J. Bae, Design and control of a wearable hand exoskeleton with force- controllable and compact actuator modules, in: 2015 IEEE International Conference on Robotics and Automation (ICRA), 2015, pp. 5596–5601.

[61] N. Popescu, D. Popescu, M. Ivanescu, D. Popescu, C. Vladu, I. Vladu, Force observer-based control for a rehabilitation hand exoskeleton system, in: 2013 9th Asian Control Conference (ASCC). IEEE, 2013, pp. 1–6.

[62] M. Ivanescu, D. Popescu, M. Nitulescu, N. Popescu, Parameter estimation techniques for a rehabilitation hand exoskeleton, in: 2014 18th International Con- ference on System Theory, Control and Computing (ICSTCC). IEEE, 2014, pp. 267–272.

[63] N. Popescu, D. Popescu, M. Ivanescu, D. Popescu, C. Vladu, C. Berceanu, M. Poboroniuc, Exoskeleton design of an intelligent haptic robotic glove, in: 2013 19th International Conference on Control Systems and Computer Science, 2013, pp. 196–202.

[64] M. Takaiwa, Wrist rehabilitation training simulator for pt using pneumatic parallel manipulator, in: 2016 IEEE International Conference on Advanced Intelligent Mechatronics (AIM). IEEE, 2016, pp. 276–281.

[65] M. Takaiwa, T. Noritsugu, Development of wrist rehabilitation equipment using pneumatic parallel manipulator, in: Proceedings of the 2005 IEEE International Conference on Robotics and Automation. IEEE, 2005, pp. 2302–2307.

[66] M. Takaiwa, T. Noritsugu, Development of wrist rehabilitation equipment using pneumatic parallel manipulator-acquisition of pt's motion and its execution for patient, in: 2009 IEEE International Conference on Rehabilitation Robotics. IEEE, 2009, pp. 34–39.

[67] M. Takaiwa, T. Noritsugu, Wrist rehabilitaion equipment using pneumatic parallel manipulator, in: 2010 World Automation Congress. IEEE, 2010, pp. 1–6.

[68] M. Saadatzi, D.C. Long, O. Celik, Torque estimation in a wrist rehabilitation robot using a nonlinear disturbance observer, in: ASME 2015 Dynamic Systems and Control Conference. American Society of Mechanical Engineers, 2015, pp. V001T18A001.

[69] M. Saadatzi, D.C. Long, O. Celik, Comparison of human-robot interaction torque estimation methods in a wrist rehabilitation exoskeleton, J. Intell. Robot. Syst. 94 (3–4), (2018) 1–17.

[70] A.U. Pehlivan, F. Sergi, M.K. O'Malley, A subject-adaptive controller for wrist robotic rehabilitation, IEEE/ASME Trans. Mechatron. 20 (3) (2015) 1338–1350.

[71] A.U. Pehlivan, D.P. Losey, M.K. O'Malley, Minimal assist-as-needed controller for upper limb robotic rehabilitation, IEEE Trans. Robot. 32 (1) (2016) 113–124.

[72] J. Niu, Q. Yang, G. Chen, R. Song, Nonlinear disturbance observer based sliding mode control of a cable-driven rehabilitation robot, in: 2017 International Conference on Rehabilitation Robotics (ICORR), 2017, pp. 664–669.

[73] J. Niu, Q. Yang, X. Wang, R. Song, Sliding mode tracking control of a wire-driven upper-limb rehabilitation robot with nonlinear disturbance observer, Front. Neurol. 8 (2017) 646.

[74] G. Chen, C.K. Chan, Z. Guo, H. Yu, A review of lower extremity assistive robotic exoskeletons in rehabilitation therapy, Crit. Rev. Biomed. Eng. 41 (4–5) (2013) 343–363.

[75] N.N. Byl, Mobility training using a bionic knee orthosis in patients in a post-stroke chronic state: a case series, J. Med. Case Rep. 6 (1) (2012) 216.

[76] P.K. Jamwal, S. Hussain, S.Q. Xie, Review on design and control aspects of ankle rehabilitation robots, Disabil. Rehabil. Assist. Technol. 10 (2) (2015) 93–101.

[77] H. Yu, S. Huang, G. Chen, Y. Pan, Z. Guo, Human–robot interaction control of rehabilitation robots with series elastic actuators, IEEE Trans. Robot. 31 (5) (2015) 1089–1100.

[78] R. Kotina, Q. Zheng, A.J. van den Bogert, Z. Gao, Active disturbance rejection control for human postural sway, in: Proceedings of the 2011 American Control Conference. IEEE, 2011, pp. 4081–4086.

[79] Y. Tsoi, S. Xie, G. Mallinson, Joint force control of parallel robot for ankle rehabilitation, in: 2009 IEEE International Conference on Control and Automation. IEEE, 2009, pp. 1856–1861.

[80] Y. Tsoi, Modelling and adaptive interaction control of a parallel robot for ankle rehabilitation, Ph.D. dissertation, ResearchSpace@ Auckland, 2011.

[81] P.K. Jamwal, S.Q. Xie, S. Hussain, J.G. Parsons, An adaptive wearable parallel robot for the treatment of ankle injuries, IEEE/ASME Trans. Mechatron. 19 (1) (2014) 64–75.

[82] Q. Ai, C. Zhu, J. Zuo, W. Meng, Q. Liu, S. Xie, M. Yang, Disturbance- estimated adaptive backstepping sliding mode control of a pneumatic muscles-driven ankle rehabilitation robot, Sensors 18 (1) (2018) 66.

[83] G. Colombo, M. Joerg, R. Schreier, V. Dietz, et al. Treadmill training of paraplegic patients using a robotic orthosis, J. Rehabil. Res. Dev. 37 (6) (2000) 693–700.

[84] S. Mohammed, W. Huo, J. Huang, H. Rifaï, Y. Amirat, Nonlinear disturbance observer based sliding mode control of a human-driven knee joint orthosis, Robot. Auton. Syst. 75 (2016) 41–49.

[85] H. Rifai, S. Mohammed, K. Djouani, Y. Amirat, Toward lower limbs functional rehabilitation through a knee-joint exoskeleton, IEEE Trans. Control Syst. Technol. 25 (2) (2016) 712–719.

[86] H. Rifai, S. Mohammed, B. Daachi, Y. Amirat, Adaptive control of a human- driven knee joint orthosis, in: 2012 IEEE International Conference on Robotics and Automation, 2012, pp. 2486–2491.

[87] M. Khamar, M. Edrisi, Designing a backstepping sliding mode controller for an assistant human knee exoskeleton based on nonlinear disturbance observer, Mechatronics 54 (2018) 121–132.

[88] J. Lu, K. Haninger, W. Chen, M. Tomizuka, Design and torque-mode control of a cable-driven rotary series elastic actuator for subject-robot interaction, in: 2015 IEEE International Conference on Advanced Intelligent Mechatronics (AIM), 2015, pp. 158–164.

[89] T. Nguyen, S.J. Allen, S.J. Phee, Direct torque control for cable conduit mechanisms for the robotic foot for footwear testing, Mechatronics 51 (2018) 137–149.

[90] D.J. Reinkensmeyer, C.T. Pang, J.A. Nessler, C.C. Painter, Web-based telerehabilitation for the upper extremity after stroke, IEEE Trans. Neural Syst. Rehabil. Eng. 10 (2) (2002) 102–108.

[91] M.J. Rosen, Telerehabilitation, NeuroRehabilitation 12 (1) (1999) 11–26.

[92] S. Ueki, H. Kawasaki, S. Ito, Y. Nishimoto, M. Abe, T. Aoki, Y. Ishigure, T. Ojika, T. Mouri, Development of a hand-assist robot with multi-degrees-of-freedom for rehabilitation therapy, IEEE/ASME Trans. Mechatron. 17 (1) (2010) 136–146.

[93] G. Burdea, V. Popescu, V. Hentz, K. Colbert, Virtual reality-based orthopedic telerehabilitation, IEEE Trans. Rehabil. Eng. 8 (3) (2000) 430–432.

[94] M.D. Duong, C. Teraoka, T. Imamura, T. Miyoshi, K. Terashima, Master-slave system with teleoperation for rehabilitation, IFAC Proc. Vol. 38 (1) (2005) 48–53.

[95] X. Wang, J. Li, A tele-rehabilitation system with bilateral haptic feedback to both the therapist and the patient via time-delay environment, in: 2013 World Haptics Conference, 2013, pp. 331–334.

[96] M. Sharifi, S. Behzadipour, H. Salarieh, M. Tavakoli, Cooperative modalities in robotic tele-rehabilitation using nonlinear bilateral impedance control, Control Eng. Pract. 67 (2017) 52–63.

[97] J. Lanini, T. Tsuji, P. Wolf, R. Riener, and D. Novak, Teleoperation of two six- degree-of-freedom arm rehabilitation exoskeletons, in *2015 IEEE International Conference on Rehabilitation Robotics (ICORR)*, 2015, pp. 514–519.

[98] A. Gupta, M.K. O'Malley, Disturbance-observer-based force estimation for haptic feedback, J. Dyn. Syst. Meas. Contr. 133 (1) (2011) 014505.

[99] R. Oboe, P. Fiorini, Issues on internet-based teleoperation, IFAC Proc. Vol. 30 (20) (1997) 591–597.

[100] E. Nuño, L. Basañez, R. Ortega, M.W. Spong, Position tracking for non-linear teleoperators with variable time delay, Int. J. Robot. Res. 28 (7) (2009) 895–910.

[101] W. Zhang, M. Tomizuka, J. Bae, Time series prediction of knee joint movement and its application to a network-based rehabilitation system, in: 2014 Amer- ican Control Conference. IEEE, 2014, pp. 4810–4815.

[102] J. Bae, K. Kong, M. Tomizuka, Control algorithms for prevention of impacts in rehabilitation systems, in: 2011 IEEE/ASME International Conference on Advanced Intelligent Mechatronics (AIM). IEEE, 2011, pp. 128–133.

[103] J. Bae, W. Zhang, M. Tomizuka, Network-based rehabilitation system for improved mobility and tele-rehabilitation, IEEE Trans. Control Syst. Technol. 21 (5) (2012) 1980–1987.

[104] K. Natori, R. Oboe, K. Ohnishi, Stability analysis and practical design procedure of time delayed control systems with communication disturbance observer, IEEE Trans. Ind. Inform. 4 (3) (2008) 185–197.

[105] K. Natori, T. Tsuji, K. Ohnishi, A. Hace, K. Jezernik, Time-delay compensation by communication disturbance observer for bilateral teleoperation under time-varying delay, IEEE Trans. Ind. Electron. 57 (3) (2009) 1050–1062.

[106] W. Zhang, M. Tomizuka, Compensation of time delay in a network-based gait rehabilitation system with a discrete-time communication disturbance observer, IFAC Proc. Vol. 46 (5) (2013) 555–562.

CHAPTER 6

Reduction in the metabolic cost of human walking gaits using quasi-passive upper body exoskeleton

Nafiseh Ebrahimi, Gautham Muthukumaran, Amir Jafari, Andrew Luo
Mechanical Engineering Department, University of Texas at San Antonio (UTSA), One UTSA Circle, San Antonio, TX, United States

1 Introduction

Humans walk in many different styles in the daily life and there are various factors that influence the energetics of walking. The energetics of human walking is already efficient but still gaining interests among the researchers due to the mass population involved in it. Dynamic walking has been a postulated method to deal with the legged locomotion which accentuates the utilization of basic dynamical models and concentrates on implementation through the span of several strides, contrary to governing a single stride, predominantly intends to comprehend or advance the relation between walking gaits and the related energy costs. There is a close interconnection between dynamic walking and passive walking. Ted McGeer as a pioneer in passive dynamic walking hypothesized the improvement in the control of walking robots via designing it to follow a gait with an oscillating system similar to the swinging action of a pendulum [1]. In the present study, we approached the upper body dynamics during walking gaits with a similar concept to design an oscillating exoskeleton which could bring about a correlation between the gait and the energetics of walking. During mid-1900, the development of motion capture framework enabled Fischer to formulate equations of motion for a mechanical model of a human walking [2]. In the 1930s, Elftman utilized a similar framework on film to finish a full inverse-dynamics of the torques required for each lower limb joint in the stride cycle [3]. Passive dynamic models were non–complex and few parameters driven to portray each model. This took into consideration a more explorative investigation of parameter space compared to that in

Powered Prostheses
http://dx.doi.org/10.1016/B978-0-12-817450-0.00006-7

complex high-parameter models. Similarly, the results were less demanding to understand, made it simpler to comprehend the standard components of the frameworks. Rather than concentrating on the stability of the joint directions within a step, passive dynamics considers stability of a cluster of steps over a time span.

Human-subject experiments based on dynamic walking are model-based as contradicted to observation-based as mentioned in Ref. [4]. Traditionally, numerous hypotheses raised in the biomechanics researches with regards to the hidden standards of human walking constructed altogether with respect to perceptions of unaffected gait analysis.

Biped locomotion in humans is an evolution from Quadrupedalism. The necessary locomotors are the lower limbs but the upper limbs move either passively or actively facilitating the movement of legs [5]. Humans are the most efficient species practicing biped locomotion as their primary walking gait (McNab 2002).

Initial clinical examinations on gaits have a tendency to disregard arm swing assuming head (H), arms (A), and trunk (T) into a HAT unit which moves as one mass altogether [6,7]. The question of how arm swing starts while walking is ambiguous but the question of why has resulted in interesting findings as per Ref. [8] research. First, arm swing during human gait reduces energetic cost by 8%. Secondly, swinging the arms during gait facilitates the movements of the legs. Studies also suggest that preventing arm swing changes the gait pattern in healthy adults affecting predominantly inter-limb coordination and walking cadence [9,10].

Throughout the years, numerous reasons have been reported, including improving stability, optimizing energy utilization, and most extreme utilization of the neuronal legacy of quadrupedal gait. Arm swing help in recovering balance after confusion and that might be embraced by ones in danger of falling [11]. Later examinations, affirmed that arm swing diminished angular momentum about the vertical axis as well as vertical ground reaction moment [12] and in particular, actual energy consumption for which ranges extend up to 8% energy utilization in contrast to folded-arm gait [13,14].

Additionally, higher order control of inter-limb coordination can be accomplished at brainstem and cortical level which leads to acceptance of patterns generated from the brain during walking. The steps of leg swing and arm swing may be due to Central Pattern Generation (CPG) from brainstem [15]. But later studies did not support or provide evidence for the presence of CPG in human for generating walking rhythm. The normal-biped

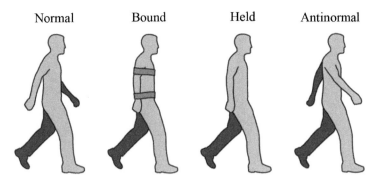

Normal   Bound   Held   Antinormal

Figure 6.1 Various walking gaits presented by Collins including Normal: arms swing in opposite phase with legs, Bound: arms physically bound to the body, Held: arms held purposely to the body; Anti-normal: arms deliberately swing with phase opposite to the normal. [19].

locomotion involves upper limbs and lower limbs swinging against each other as shown in the first feature of Fig. 6.1.

Various estimation techniques have been utilized, extending from basic photographic methods to more refined three-dimensional sensing of markers set at known positions on the body. People swing their arms when they walk as a consequence of both passive inclinations and muscular forces. Both passive inclinations and muscular forces grant the motion of the arms however, their relative commitments have stays debatable (Figs. 6.2 and 6.3)

The noteworthy portion of the gait examination is the estimation of the kinetics applied to the body. Vertical and horizontal ground reaction forces applied to the feet and their purpose of utilization can be measured straightforwardly, and are helpful in distinguishing the net forces and moments applied to the body. Another imperative aspect of gait analysis is energy utilization, which gives a measure of the energy required to do different exercises like walking, running, and so on.

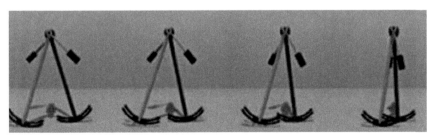

Figure 6.2 Simulations of walking gaits by Ref. [23].

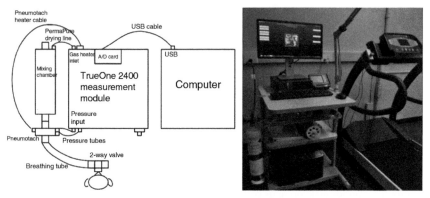

Figure 6.3 Block diagram and actual set up of the experimental apparatus.

Until the mid-twentieth century early Bio-mechanists Morton, 1952, Gerdy, 1829, and Weber, 1836 proposed that the arms may swing absolutely because of the muscle movement of the shoulders during the walk similar to a pendulum [16]. In Ref. [17] and [18], it was demonstrated that joint torques including those emerging from muscle contraction occupy a part of the movement of the arms. Their dynamic investigations have brought about a span of resultant joint moments at the shoulder, from 3.8 Nm to 7.5 Nm and to a greatest of 12 Nm. It has also been assumed that arm swinging might be a transformative relic from quadrupedalism which no practical value has been reported.

The investigations in Ref. [19] allowed releasing the arms during walking of a three-dimensional passive dynamic walking biped robot. A few ways of arm swing were created by Collins including—Normal: each arm waves in stage with the leg on the inverse side; Bound: the arms are mechanically held attached to the body restraining motion; Held: the arms purposely held to the body; Antinormal: each arm deliberately swings in accordance with the leg on a similar side.

In Ref. [20], it was performed a number of experiments using motion capture system with the different walking gaits to find out the fact that the absence of arm swing increases metabolic cost and also arm swing is complemented passively. These findings were earlier affirmed by electromyography discoveries in Ref. [21, 22]. Later investigations contemplated that such dynamic shoulder torques are just trivial and then suggested that arm swinging might be to a greatly passive.

One of the key issues in the biomechanics has been the computation of forces and torques produced inside the muscles and joints during human movements. The estimation of such data is huge and time consuming for the experimental procedures. It was hypothesized in Ref. [21] that the high metabolic expenses of motion without swinging the arms are either due to the ground moment around the body vertical axis which should be investigated and could be owing to the displacement of the center of mass.

Arm swinging mode fundamentally influenced vertical angular momentum and ground reaction moments in the simulations created in Ref. [23]. The peak of vertical angular momentum and the peak of vertical ground reaction moments increased as the arms swing went from normal to bound to antinormal. The proposed speculation is that arm movements could utilize microscopic metabolic cost, requiring solid action just to begin the arms in movements.

The addressed literature seems to give clashing conclusions with respect to the estimation of mechanical model examination in evaluating metabolic cost, however, gave some helpful understanding into the sort of correlations that ought to be fulfilled to conduct the current research.

The first and foremost requirement of this study was to understand the energetics of different human walking gaits. The walking pattern varies from one person to another. The amplitude and frequency of person walking varies at different speed levels and also depends on practice exhibited for a long period of time. Aforementioned studies inferred metabolic energy cost might be a solid determinant in the kinematic patterns of walking. A framework is required to control the vertical orientation of the lateral and dorsal velocity of the model.

Steve Collins and Greg Sawicki found an approach to make people walk more efficiently [24]. They equipped a lower body unpowered exoskeleton to demonstrate a diminishment of 7% on the aggregate metabolic cost for human walking. This inference gave us an insight to find ways of reducing metabolic cost for walking with the use of exoskeletons. In the present paper, In contrast to the lower body exoskeleton, we concentrated our research toward reducing the metabolic rate on the upper body.

Interestingly, researchers who had earlier worked on reducing metabolic cost looked down to the lower limbs. However, our research is oriented toward the upper body contribution to the lower body locomotion. It is a study on a combined effort by upper and lower limbs in accomplishing the locomotion of humans through biped walking. To bring a consensus that replacing arm swing with something else could cause a reduction in metabolic cost, we accomplished some experiments with different subjects and their different gaits to study what influences could absence of arm swing have on human walking patterns which consequently might result in some promising improvements in the life of some people who look for rehabilitation of their normal gait after an accident and arm amputees and consequently consume more energy during walking. Therefore, in the present study we analyzed the necessity of arm swing for efficient human walking gait and studied the association between arm swing and metabolic cost of the walking/running. Then, we studied some potential efficient means to cause the reduction in the metabolic cost and finally investigate the hypothesis that if the use of upper body exoskeletons could affect the metabolic cost of the walking/running.

For these purposes, a group of six subjects was asked to walk in four different gaits on a treadmill to record and analyze their energy consumption respiratory parameters. The idea was to enable subjects to walk with a custom designed upper body exoskeleton. Accordingly, a quasi-passive exoskeleton which suits all of the subjects was designed and fabricated to perform experiments to study the reduction of metabolic cost in folded-arm walking gait.

2 Methodology

The open-circuit strategy is the most broadly utilized part of established instruments for measuring energy consumption. Caloric consumption measured by indirect calorimetry combined with the double-marked water procedure presented the idea of physical activity metabolic cost, which added to resting energy use brings about aggregate everyday metabolic consumption. Compact modular and handheld gadgets have been brought into the market, together with comparative innovation for assessing activity energy requirement.

Indirect calorimetry is an ideal method for measuring gas flows which primarily includes volumes and concentrations of O_2 and CO_2 inspired and expired, respectively into the human body. The system determines a measurement of oxygen consumption and CO_2 production of a person while performing an activity. The technique is noninvasive and accurate as well as easy to use. The equipment used in this method was also known as a metabolic cart.

This strategy utilizes a constant estimation of O_2 consumption (VO_2) and CO_2 production (VCO_2). It determines caloric needs and calculates substrate oxidation by figuring respiratory Quotient (CO_2 production/O_2 consumption) and resting energy expenditure (REE).

There are different parameters measured and calculated by the metabolic cost estimation experiment. The parameters are measured by allowing subjects to walk on the treadmill at various speed levels and monitoring their metabolic activity rates with the calorimeter and heart beat rate sensor.

The parameters measured and calculated include:

- VO_2 - Measured amount of injected oxygen during exercise (L/min)
- VCO_2 - Volume of CO_2 exhaled during exercise (L/min)
- VE - Amount of air moved in and out of the lungs per minute (L/min)
- RER - Respiratory exchange ratio, VCO_2/VO_2
- RR - Respiration rate (BPM)
- FEO_2 – The proportion of O_2 in the exhaled air (%)
- $FECO_2$ - The proportion of CO_2 in the exhaled air (%)
- MET - Metabolic equivalent of a task or multiples of an individual resting oxygen uptakes (3.5 mL O_2/min.kg)

2.1 Testing apparatus and procedure

The Parvo Medics TrueOne® 2400 is an integrated metabolic measurement system for maximal O_2 consumption testing and indirect calorimetry assessment. The TrueOne® 2400's analyzer module, with paramagnetic oxygen and infrared carbon dioxide analyzers, is designed to be accurate, reliable, and easy to use. The unique flowmeter calibration algorithm, utilizing an image reconstruction technique, corrects the nonlinearity of the pneumatic system, and provides highly accurate flow measurement. Subjects were allowed to walk on the treadmill at different speeds to determine the ideal speed of walking. Four-speed levels were chosen to experiment with each subject. Heartbeat sensor is worn by the subject to monitor and record the heartbeat rate during the walking. An initial warm up period of 2–3 minutes is provided for the subject to get accustomed to the system. Two speed levels one above and below the ideal walking speed as well as one jogging speed were selected to perform the experiment. The subject fulfilled each gait for a 2–3 minutes period allowing 4–6 readings per gait. The total duration of the experiment was 10–15 minutes comprising of all four gaits together.

The experiment is rerun to obtain 64 readings per person. A total of 6 people including 3 males and 3 females of weight ranging from 45 to 85 kg served as the test subjects for the experiments. Four different gaits consist of namely—Normal which the subject walks normally with arm swing in opposite phase with leg swing; Antinormal that the subject walks with arm swing exactly in phase with leg swing; Folded arm that the subject walks

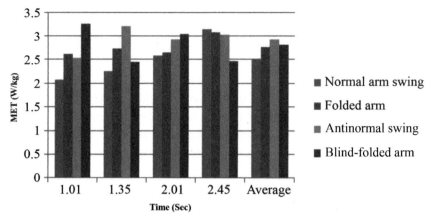

Figure 6.4 MET (Metabolic equivalent for a task) comparison for different gaits at speed 2 mph.

with arms folded while walking on the treadmill; And finally, Blind folded which the subject is allowed to walk with eyes closed and arms folded while walking on the treadmill.

Metabolic equivalent for a task (MET) is the key parameter of comparison as it determines the amount of energy spent during a particular exercise such as walking and jogging in our experiments. Applying the acquired data from metabolic cart, the metabolic equivalent (E) were calculated and tabulated through using Brockway equation [equation (1)] [25].

$$E = 16.58kJ/L(O_2) + 4.51kJ/L(CO_2) - 5.90kJ/g(UreaNitrogen) \qquad (1)$$

The calculated MET results for speeds 2 and 3 MPH are presented in Fig. 6.4 and Fig. 6.5 which indicate that folded arm, antinormal, and blind-folded modes of arm swinging had a significant increase in MET values as compared to normal one consequently resulted in a significant net increase in metabolic rates. The increase in antinormal and blind folded can be inferred easily as it is against natural walking harmonics of humans. But the increase in the energy consumption level for folded–arm gait was alarming as well as interesting. This is because humans will not always be able to move their arms while walking. Other ways this means that people tend to spend more energy while carrying a load on their arms or if their arm swing is not present because of the absence of arms or rehabilitation gait.

The presented results brought us to a conclusion that arm swing is required for efficient walking and what happens if a person cannot move

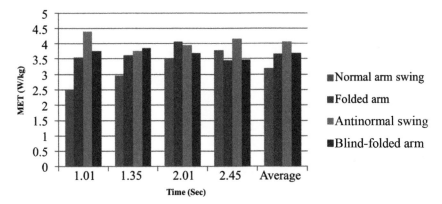

Figure 6.5 MET (Metabolic equivalent for a task) comparison for different gaits at speed 3 mph.

his arms while walking seemed to be an unanswered question from this aspect. The conclusion has a good agreement with Collins et al. findings. In addition, the increase in the metabolic cost due to the absence of arm swing gave a consensus to reduce MET with upper limb motion. This gave us the insight and motivation to exert an exoskeleton which could change the energetics of human walking and other activities.

In the following the design and fabrication of the upper body exoskeleton is explained. The exoskeleton was applied to the subjects in order to perform similar tests to measure respiratory parameters and calculate the MET as a criterion of energy consumption for different arm swing modes while walking gate.

2.1.1 Upper body passive exoskeleton fabrication and experimental results

In the designed exoskeleton, the oscillation of a pendulum served as stimuli for the design. A mechanical design which could bring an oscillation like a human hand movement was the first ideal choice for the exoskeleton design. The movement of the exoskeleton arm should be synchronized with the leg swing and should not be faster or slower than the same. The initial design was entirely based on the finding that arm swing is passive and setting up conditions for movement of the pendulum structure around the shoulder can be the appropriate way to do it.

Artec Eva three-dimensional scanner was used for scanning the subject's shoulder muscle to achieve a three-dimensional scan output in order to be

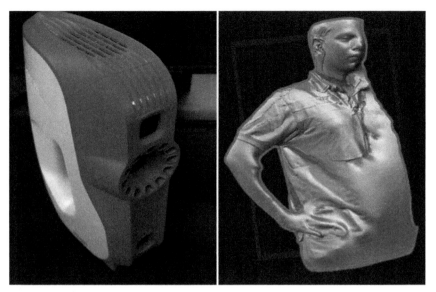

Figure 6.6 Artec Eva three-dimensional scanner and the three-dimensional shoulder scan of the subject.

used in designing the exoskeleton (Fig. 6.6). The scan was acquired to fit in as a mate for the shoulder sleeve. The shoulder scan of the subject was obtained to design an arm sleeve specifically for him to prepare a custom-made exoskeleton. The pivot point was the shoulder joint about which the movement of arms was actuated for mimicking the motion of a pendulum (Fig. 6.7).

Two pendulums-like arms each of 15cm length were designed and 3D printed to oscillate about the vertical axis. These pendulums were made to hold together with variable possible lengths to adjust the natural swing frequency of oscillation. A very small amount of weight was added to the end of the lower pendulum structure to sustain the oscillation for a longer period of time without much actuation. An internal pendulum structure which could fit in between the other two pendulums plays a crucial role to vary the length of the pendulum as per the frequency requirement.

The amplitude of the swing was also varied with the aid of weight at the lower part of the pendulum. The initial setting was very critical for this design to ensure the desired motion of the pendulum.

The same set of experiments was performed by the subject wearing the exoskeleton around the shoulder. The initial setting was made with the

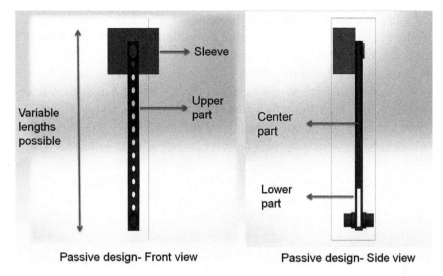

Figure 6.7 Schematic views of the fully passive exoskeleton design.

pendulum structures held against each other in opposite directions to bring normal swing of arms while allowing the subject to walk in folded-arm gait. Different lengths which were equal to, more and less than the subject's arm length were tested and found that none of the results were satisfactory as depicted in Fig. 6.8.

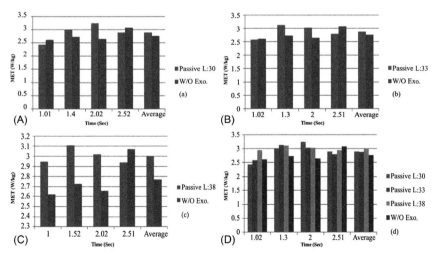

Figure 6.8 Passive exoskeleton MET experiment results for arm folded walking gait compares applying exoskeleton with the arm length of (A) 30 mm, (B) 33 mm and (C) 38 mm, versus not using exoskeleton (D).

The inference was that the MET value increased by a great value and increased further more as the speed of the walking grew. The value of MET ranged from 2.64 to 3.01 W/kg without the exoskeleton and with the exoskeleton, it extended from 2.85 to 3.22 W/kg. Across all possible lengths, the average value without exoskeleton was 2.71 W/kg much less than the average found with the fully passive exoskeleton which was 2.95 W/kg. Therefore, testing with passive upper body exoskeleton was not very effective even with performing with different lengths at various speed levels.

Applying fully passive exoskeleton despite the negative impacts on the MET values has a few significant findings out of the experiments:

- Projection of the external weight onto the body caused interesting changes as the subject felt it hard to walk with it and also might be one of the reasons for the negative results obtained.
- The swing of pendulum was not perfectly symmetric about vertical axis of the shoulder of the subject and the amplitude tends to reduce once the motion lasts for a particular period of time.
- Smooth exoskeleton arm swing was not obtained as it found it hard to sustain the swing with the desired frequency for a longer period of time.
- Quasi-passive upper body exoskeleton was another proposed alternative suggested since the main idea of mimicking the arm swing did not occur as expected. Possibly this was the main reason which did not bring a reduction in the level of MET measured with fully passive exoskeleton.

2.1.2 Upper body quasi-passive exoskeleton fabrication and experimental results

Two major features which should be considered in the design of the quasi-passive exoskeleton were to make it continuously oscillate with stable frequency and amplitude and another was to synchronize the swing of the arm and leg perfectly complementing each other. The idea to use Dynamixel motors to actuate the pendulum and mimic human arm swing was the best option available and it helped out. The revised quasi-passive exoskeleton design embedding the Dynamixel motors is illustrated in the Fig. 6.9.

An exoskeleton assembly of 24 parts (12 parts each) was designed, 3D printed with the same material and then integrated with a shoulder brace. The role of the shoulder brace was to provide the pivot point for rotation of the exoskeleton arms about the vertical axis. It helped by providing comfort to the subject and also holding the exoskeleton intact closer to the shoulder of the subject (Fig. 6.10).

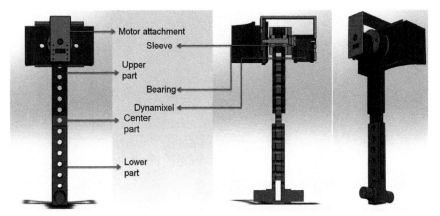

Figure 6.9 Schematic front, side, and isometric views of the quasi-passive exoskeleton design.

Figure 6.10 Shoulder brace, Dynamixel AX-12A motor and OpenCM 9.04 microcontroller used in the fabrication of quasi-passive exoskeleton.

The swing of the exoskeleton arms was controlled by programming the open source IDE available. OpenCM 9.04 is a microcontroller board based on 32bit ARM Cortex-M3. Fig. 6.11 depicts the shoulder brace, Dynamixel motor and OpenCM 9.04 microcontroller.

The control of Dynamixel motors (Dynamixel AX-12A) was quite straightforward because it has a microprocessor which could be commanded with the OpenCM 9.04. This assembly used a power source of 12V SMPS to power the exoskeleton to mimic the motion of arms. The entire idea was to replace human arm swing with the exoskeleton arm swing during folded-arm gait which could practically be seen as a subject walking with load or no arms. For this entire set of experiments just

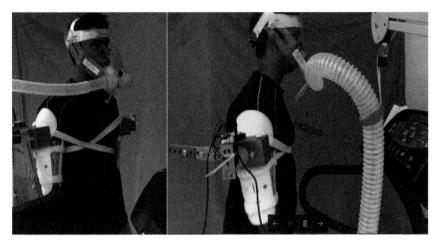

Figure 6.11 Subject performing the experiments.

one subject served. Fig. 6.11 shows the subject wearing the quasi-passive exoskeleton and performing the experiments after the test set up. Critical analysis was accomplished for the symmetric oscillation about the vertical axis to achieve perfect mimic of the human arm swing.

The parts were all assembled using screws in different lengths also making the design comply with various lengths as expected for the natural swing frequency of the exoskeleton arms. The frequency of arm swing of the subject for different speed levels was studied and the exoskeleton was programmed to swing with exactly the same frequency for each speed level. With a change in each speed level the frequency was taken for a toss and took much time to mimic the frequency with the exoskeleton. The amplitude was also varied for different ranges to find out ideal range which could serve the purpose.

The metabolic cost estimation experiment was performed in the ideal speed level of the subject wearing the quasi-passive exoskeleton in folded-arm gait. The experiment was repeated for different speed levels for the same gait. Although the task seemed to be tough regarding the time and experience it was easy to fine tune the swing of the arm synchronize with the leg swing of the subject. The experiment results applying quasi-passive exoskeleton performed in speed levels of 2 and 3 MPH depicted in Fig. 6.12.

Unlike the previous experiment using the fully passive exoskeleton, modifying the exoskeleton by utilization of the motors to synchronize the arm swing frequency with the leg swing one and also precise initialization of the arm swing motion in the quasi-passive version led to promising results in metabolic cost reduction during folded arm-walking gait.

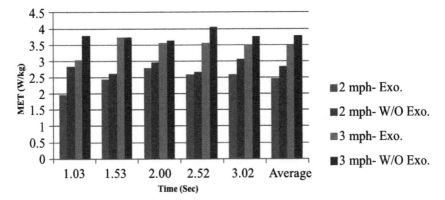

Figure 6.12 MET comparison for walking speed level 2 and 3 MPH performed with and without exoskeleton.

Table 6.1 MET reduction percentage for various walking speed levels using upper body quasi-passive exoskeleton.

Speed (MPH)	2	3	4
Reduction in MET (%)	13	9	10.3

Table 6.1 shows the MET reduction percentage obtained by the use of quasi-passive exoskeleton in various walking speeds.

The range of reduction in MET for various speed levels was from 8.8% to 13.2% for the folded-arm gait applying the upper body quasi-passive exoskeleton which makes a huge difference particularly in long walking time spans as the reduction was seen to be consistent for a period of 20–40 minutes.

The results confirm the hypothesis assumed earlier that balancing the vertical moments created by arm swing with the one made by leg swing could reduce the total metabolic cost of the human walking. Therefore, the application of the quasi passive upper body exoskeleton has a positive effect on the human walking efficiency.

3 Discussion and conclusion

We hypothesized that the contribution of arm swing to the energetics of human walking can be estimated and replacing arm swing with a similar external motion could bring a reduction in the level of energy consumption. Our results obtained with the use of upper body quasi-passive exoskeleton confirm the hypothesis and also validates an estimation of 9%–12% reduction in the metabolic cost equivalent for human walking. Further exploration of this domain would yield interesting facts and by further advancements, the reduction in human energy consumption for performing a task could be easily varied and various means of reducing the walking energy cost might be found. This area of study is related to each and every human being, so an infinitesimal positive change in the life of humans can bring a large-scale difference in a long. Army soldiers and labors who tend to carry heavy loads and machinery would benefit from this type of exoskeleton. Although this is a primitive design, a more advanced exoskeleton suit which can be easily worn could bring a remarkable difference in their lives. Another significant group who might find these exoskeletons beneficial are people who live with gait abnormality and amputee rehabilitation. People who try to retrieve their original walking gait and people who suffered from arm amputation can use this type of wearable exoskeleton to gain back their original gait. In addition, more researches as to make this type of exoskeletons easily wearable could leads to a huge industry of commercial exoskeletons.

The studies focused on assessing energy requirements for various exercises such as locomotion carrying a particular weight, backpack-loaded and front-loaded might be possible potential researches all with the basis of quest for amendment. Moreover, the necessity makes it feasible and brings much technical and technological advancement. Some of the future scopes of study related to this research include:

- Make the exoskeleton as a part of suit which can be easily portable.
- More reliable and easy to wear set-up would be the next level of extension required.
- Compact to wear custom made design for arm amputees and army soldiers.
- Study the metabolic cost for lower limb amputees to check for reduction in their metabolic cost.
- Interesting results can be obtained by combined use of upper and lower body exoskeletons
- The use of variable stiffness springs to store energy which could bring very significant updates on the energy storing capacity of exoskeletons.

References

[1] N. López Rodríguez. Extensión de la extremidad superior del robot humanoide Poppy, (2018).
[2] Fischer, R., 2000. Chapter 2: Motion Capture Process and Systems, Natick, MA: AK Peters, US, 58.

[3] H. Elftman, Forces and energy changes in the leg during walking, Am J Physiol Leg Content 125 (1939) 339–356.

[4] M. Kubo, R.C. Wagenaar, E. Saltzman, K.G. Holt, Biomechanical mechanism for transitions in phase and frequency of arm and leg swing during walking, Biol Cybern 91 (2004) 91–98.

[5] R.M. Alexander. Principles of animal locomotion. Princeton University Press, 2003.

[6] J.L. Hudson, Coordination of segments in the vertical jump, Med Sci Sports Exerc 18 (1986) 242–251.

[7] M. Kubo, B. Ulrich, Coordination of pelvis-HAT (head, arms and trunk) in anterior–posterior and medio-lateral directions during treadmill gait in preadolescents with/without Down syndrome, Gait Posture 23 (2006) 512–518.

[8] P. Meyns, S.M. Bruijn, J. Duysens, The how and why of arm swing during human walking, Gait Posture 38 (2013) 555–562.

[9] B.R. Umberger, Effects of suppressing arm swing on kinematics, kinetics, and energetics of human walking, J Biomech 41 (2008) 2575–2580.

[10] H. Pontzer, J.H. Holloway, J.H. Holloway, D.A. Raichlen, D.E. Lieberman, Control and function of arm swing in human walking and running, J Exp Biol 212 (2009) 523–534.

[11] A. Lees, J. Vanrenterghem, D.D. Clercq, Understanding how an arm swing enhances performance in the vertical jump, J Biomech 37 (2004) 1929–1940.

[12] J. Park, Synthesis of natural arm swing motion in human bipedal walking, J Biomech 41 (2008) 1417–1426.

[13] J.H. Carr, A.M. Gentile, The effect of arm movement on the biomechanics of standing up, Hum Mov Sci 13 (1994) 175–193.

[14] C.J. Arellano, R. Kram, The effects of step width and arm swing on energetic cost and lateral balance during running, J Biomech 44 (2011) 1291–1295.

[15] J. Duysens, H.W.A.A. Van de Crommert, Neural control of locomotion; part 1: the central pattern generator from cats to humans, Gait Posture 7 (1998) 131–141.

[16] P. Allard, R. Lachance, R. Aissaoui, H. Sadeghi, M. Duhaime, Able-bodied gait in men and women, Three-dimensional Anal Human Locomotion, Wiley, (1997) pp. 307–334.

[17] H. Elftman, The force exerted by the ground in walking, Arbeitsphysiologie 10 (1939) 485–491.

[18] R.N. Hinrichs, Whole Body Movement: Coordination of Arms and Legs in Walking and Running, in: J.M. Winters, S.L.-Y. Woo (Eds.), Multiple Muscle Systems: Biomechanics and Movement Organization, Springer New York, New York, NY, 1990, pp. 694–705.

[19] S.H. Collins, P.G. Adamczyk, A.D. Kuo, Dynamic arm swinging in human walking, Proc R Soc Lond B Biol Sci 276 (2009) 3679–3688, rspb20090664.

[20] S.H. Collins. Dynamic Walking Principles Applied to Human Gait. PhD dissertation, 2008, p. 116.

[21] F. Buchthal, M.L. Fernandez-Ballesteros, Electromyographic study of the muscles of the upper arm and shoulder during walking in patients with Parkinson's disease, Brain J Neurol 88 (1965) 875–896.

[22] R.E. Hosue, Upper-extremity muscular activity at different cadences and inclines during normal gait, Phys Ther 49 (1969) 963–972.

[23] S.H. Collins, M. Wisse, A. Ruina, A three-dimensional passive-dynamic walking robot with two legs and knees, Int J Robot Res 20 (2001) 607–615.

[24] S.H. Collins, M.B. Wiggin, G.S. Sawicki, Reducing the energy cost of human walking using an unpowered exoskeleton, Nature 522 (2015) 212–215.

[25] J.M. Brockway, Derivation of formulae used to calculate energy expenditure in man, Hum Nutr Clin Nutr 41 (1987) 463–471.

CHAPTER 7

Neural control in prostheses and exoskeletons

Maziar Sharbafi[a], Amirreza Naseri[b], André Seyfarth[a], Martin Grimmer[a]
[a]Lauflabor Locomotion Lab, Institute of Sport Science, Centre for Cognitive Science, Technische Universität Darmstadt, Darmstadt, Germany
[b]Mechanical Engineering School, Tarbiat Modares University of Tehran, Tehran, Iran

1 Overview

A major challenge in the design process of a powered prosthesis for the lower limb is the development of a robust and versatile control approach. Based on the generalized control framework from Tucker et al. [1], prostheses and exoskeleton control includes components such as the environment, the user, the wearable device, and the hierarchical controller. Additionally, physical and sensorial interactions exist between these components. For movement control to be coordinated between these contributing components a hierarchical controller with information about the user state, the user intention, the wearable device state, and the state of environment is required. In return, the controller can provide artificial sensory feedback to the user and the required commands to the wearable device. While humans can control their limbs in a robust and volitional manner, powered lower limb prosthetics must be able to adapt to different terrains and movements. Furthermore, the actions of the powered prostheses can only be guided by the observed state of the residual limb, rather than initiation of movement by itself. We believe that neural control approaches have the ability to overcome such functional limitations of current designs. In this regard, neural control can cover two areas of the control scheme presented by Tucker et al. [1]. First, the term neural control can be used with respect to the type of input signal. Second, the term neural control can be used to describe the approach for the high or mid level control layer. Neuromuscular control schemes include the infrastructure that characterizes the synthesis of the movement, whereas current approaches primarily parameterize observed behavior (e.g., torque angle characteristics). With the advantage of characterizing movement synthesis, neural control approaches can predict human movement to make the controller more adaptable. Thus the controller would be able to automatically adjust movements to the user needs (e.g., stability, efficiency, performance) and to respond to unexpected conditions, for example, changes in environment or body properties.

In the following section, we present the available concepts for neural signals as control input (Section 2) and neural control approaches for high or mid level control layer. While the input signal section has a strong focus on electromyograpy (EMG), the control layer section focuses on the following two areas. First, neuromusuclar model-based methods are introduced (Section 3.1). Second, template-based approaches are presented, which can be described as a simplified abstraction of the neuromuscular models. We will provide a detailed explanation of such a new approach by describing an implementation on an ankle prosthesis (Section 4).

2 Sensory signals for control

Sensory signals are a fundamental requirement for controlling a powered prosthesis for the lower limb. We categorized the signals that are exchanged within the control framework presented by Tucker et al. [1] into three groups: physical, physiological, and neural signals. The following paragraphs summarize examples for the use of sensory feedback with respect to these groups with a focus on neural signals. For those who are interested in additional sensory signals for powered prostheses for the lower limb, please refer to the following reviews [2–11].

2.1 Physical signals

Kinematic measures can be determined using sensors such as inertial measurement units (IMU), potentiometers, or joint angle encoders. The Sparky ankle used a gyro sensor to determine the shank angle and used angular velocity to determine gait velocity and gait percent [12]. A similar concept based on the thigh was used in the Ref. [13]. An IMU was also used in the Bionic anklefoot prosthesis [14] and the Vanderbilt Transtibial Prosthesis [15] to switch between control states. Aside from determining segment angles with IMUs, angle encoders were used to determine the knee or ankle joint angles [14,16,15]. Potentiometers were used to determine the joint angle in the tethered Vanderbilt knee and ankle prosthesis [17] and in a tethered prosthetic simulator [18]. To determine the toe angle, a linear potentiometer was used to determine the force of the toe-spring for the

Pantoe 1 [19]. Similar to the application in lower limb prosthetics, IMUs, goniometers, gyroscopes, and accelerometers were equally used as wearable sensors in gait analysis [20].

Strain gauges have been used to directly determine joint torque based on the measured actuator forces for the Bionic anklefoot prosthesis from MIT [21].

Force sensors were installed in series with the ball screws of a powered prosthetic leg to measure the force applied to the lever arm at the knee and the ankle and to determine the joint torque [13]. Joint torque was indirectly determined for a powered prosthetic ankle by the spring stiffness, an ankle angle encoder, and a motor encoder [22]. In Ref. [15] ankle joint torque and power were estimated based on measured motor current, known transmission characteristics, and measured joint angular velocity. The tethered Vanderbilt knee and ankle prosthesis [17] included a load cell in series with the motor for force-based control. Furthermore, it used a strain-based moment sensor to measure the moment between the socket and the prosthesis. Two additional strain gauges were used to measure the ground reaction forces (GRFs) of the forefoot and the heel. The AMP-Foot 2.0 includes force sensing resistors at the toe and the heel to determine ground contact as well as strain gauges at two springs to determine spring loading [23]. Contact sensors at the heel and the toe were also used for the Pantoe 1 prosthetic foot [19]. In addition, this prosthetic foot used a force sensor to measure the interaction forces between the user and the prosthetic foot. In contrast to most applications that have used electrical motors, pressure sensors were used in the pleated pneumatic artificial muscles of a tethered prosthetic foot [24] and in the pneumatic units of a knee [25].

Spatio-temporal measures were determined in the control approach used for treadmill-based experiments with the Walk-Run ankle [16]. In the Walk-Run ankle the angular velocity of the shank was used to determine the beginning of the gait cycle and the time of the last gait cycle to predict the gait percent for the following stride.

In addition to sensors that provide information of the human movement or of the state of the prosthesis, it is possible to directly collect information from the environment. A laser distance sensor in combination with an IMU was used [26] to identify the movement state and the upcoming terrain. Improved detection performance was achieved when fixing the laser at the waist, compared to the shank. For the waist fixation, 157 out of 160 tested terrain transitions were accurately recognized 300 ms to 2870 ms before the user switched the negotiated terrains. A laser in combination with an IMU at the waist was also used [27] to recognize terrain. Terrain detection accuracy was above 98% and terrain transitions were detected at least 0.5 s before the time required to switch the prosthesis control mode. Detection of the environment also improved the accuracy and the reliability of the locomotion mode recognition compared to the recognition without environmental information. In Ref. [28] four infrared sensors were mounted underneath the foot to detect objects placed up to 0.3 m away. The information was used to determine foot orientation with respect to the ground. Depth sensing was performed with the help of a Microsoft Kinect to evaluate the feasibility of detecting a staircase and the properties of a staircase [29]. Stair detection achieved an accuracy of 98.8%. This approach was able to accurately estimate the distance, the angle of intersection, the number of steps, the stair height, and the stair depth. Visual information from a camera at the shank was extracted [30] to gather information about the type of terrain and the inclination. The authors showed that six types of terrain could be predicted with an average accuracy of 86%. Furthermore, it was found that the error obtained by the proposed method for calculating the inclination was reasonably low.

2.2 Physiological signals

Mechanomyography (MMG), which measures the sound of the muscles, and sonomyography (SMG), which measures muscle thickness, were explored to characterize human muscle behavior and to control wearable robots [31,32]. So far MMG and SMG were mainly investigated for the use in upper limb prostheses. Ultrasound was also used to explore control principles in upper limb prostheses [33]. Based on Ref. [34], ultrasound imaging for sensing mechanical deformation can overcome several limitations of surface EMG, including EMG's inability to differentiate between deep contiguous muscle compartments, low signal-to-noise ratio, and the lack of a robust graded signals. Akhlaghi et al. showed that ultrasound imaging can be used as a robust muscle-computer interface [34]. In this experiment, the real-time image-based control of a virtual hand showed an average classification accuracy of 92% for distinguishing hand movements based on ultrasound imaging of forearm muscles. Sierra et al. showed that a linear relationship exists between ultrasound image features of the human forearm and the hand kinematic configuration as well as fingertip forces [35].

Metabolic cost was measured and used as a feedback parameter for humanin-the-loop optimization of wearable lower limb robots [36,37]. Heart rate, breathing frequency, and/or minute ventilation were recommended to be used as a feedback parameters by Ingraham et al. [38].

2.3 Neural signals

Navarro et al. summarized that the peripheral nervous system could be interfaced by chemical, mechanical, magnetic, and electrical solutions [39]. Based on their review, electrodes are most commonly used for either potential measurement or for current stimulation and voltage stimulation. With an increased level of electrode invasivity, the selectivity of signals for neural or muscular interfaces can be improved.

As, noninvasive surface electromyographic signals of residual muscles were used to interface with prosthetic limbs. While this technique is commonly used to control upper limb prostheses [40,33], no commercial powered lower limb prostheses currently uses the signal. In research, the idea has been explored for approximately 50 years to either use the signal to determine the user state, to determine the user intention, or to volitionally control the prosthetic limb volitional. In 1972 Horn et al. developed a design with a clutch that locks a prosthetic knee joint based on voluntary activity of residual thigh muscles [41]. In 1975 Dyck et al. presented an approach where residual thigh muscles were used to voluntarily vary the resistance of a prosthetic knee by controlling the flow around a hydraulic cylinder [42]. In Ref. [43] a method was presented to control the Vanderbilt powered knee during nonweight-bearing activity, such as sitting, by quadriceps and hamsting surface EMG of the residual limb. Nonweight-bearing real-time control based on myoelectric signals was also performed by the same device [44]. It is demonstrated that this control concept could accurately classify sagittal plane motions of the knee and ankle based on surface EMG of nine thigh muscles (semitendinosus, sartorius, tensor fasciae latae, adductor magnus, gracilis, vastus medialis, rectus femoris, vastus lateralis, biceps femoris) [44]. EMG was combined with pattern recognition [45] to identify user locomotion modes for different terrains in amputees and nonamputees. The results showed reliable classification with 10 electrodes around the knee joint for seven tested modes. Surface EMG at the anterior and posterior thigh was used to control a powered knee during arbitrary motion commands and level walking experiments [46]. The controller used an active component to reflect the active effort of the user and a reactive component to model the joint impedance. Preliminary data with a powered ankle demonstrated that a transtibial amputee is able to volitionally adapt calf EMG activity to modulate toe-off angle, net ankle work, and peak power across a broad range of walking speeds [47]. A control strategy for a prosthesis of a transfemoral amputee was developed by Guo et al. [48]. They used surface EMG of five muscles (rectus femoris, biceps femoris, semitendinosus, gastrocnemius medialis, soleus) and a support vector machine to identify the movement mode [48]. Residual thigh muscle EMG was used to enable a transfemoral amputee to directly control the knee torque of a powered prosthetic knee during stair climbing [49]. While the amputee had to effectively modulate the knee power output during stance, only modest modulation was required during swing. Instead of supporting the entire gait cycle, volitional control of knee flexion and extension during swing was explored [50] with the help of thigh muscle EMG. Gastrocnemius, soleus, and tibialis EMG were used to predict the movement of a powered prosthetic ankle [51]. And two implementation approaches were compared, a biomimetic EMG-controller and a neural network approach. While both controllers were able to predict the desired movement, the biomimetic EMG-controller demonstrated a smoother and more natural behavior. EMG of the knee flexors and extensors was used [52] to estimate the knee angle. When combining the estimation with signals from a gyroscope, it was possible to improve the knee angle estimation performance in the presence of artifacts.

Tucker et al. identified different options for transforming EMG signals to useable joint torque outputs [1]. One could use the EMG activity directly and scale it to the joint output torque. EMG can be also mapped to the desired joint state [53], or to the setpoint angle or stiffness of an impedance control law. It is also possible to use a neuromuscular model with the EMG as an input and the joint torques as an output.

While researchers have tried a variety of approaches of using EMG signals for improved lower limb prosthetic control, the use of EMG involves a lot of challenges. Tucker et al. summarized [1] that surface EMG is sensitive to many conditions, such as electrode–skin conductivity, electrode misalignment, motion artifacts, muscle fatigue, and crosstalk between nearby muscles [54–56]. An additional problem is the detection of movement onset [57]. In addition to these challenges, the use of EMG can require several steps for the analysis. Zecca et al. summarized that signal conditioning and preprocessing, feature extraction, dimensionality reduction, pattern recognition, and offline and online learning are used [58]. Furthermore, calibration of the EMG signal, for example by normalizing it to the maximum voluntary contraction [47], is required.

To improve EMG-based volitional control, training seems to be beneficial as Alcaide-Aguirre et al. [59] demonstrated that training can substantially improve volitional control for amputees with poor volitional control of their residual muscles. To improve the classification of the user state or the user intention, neuromuscularmechanical fusion has shown promise. Using this approach, the fusion of signals outperformed methods that used only EMG signals or mechanical information, when classifying locomotion-modes based on gluteal and residual thigh muscles, GRFs, and moments from the prosthetic pylon [60].

Another technique that has been explored to control wearable robotics [61,62] is electroencephalography (EEG). Based on the review of Al et al. [61], this technique was primarily used for exoskeletons and upper limb prostheses. One case study explored the feasibility of locking and unlocking a prosthetic knee directly by using a braincomputer interface (BCI) [63]. While the subject was always able to unlock the knee for stand to sit transitions they could not always unlock the knee during walking. Based on the review [61] and the case study [63], the usability, the reliability, and the translational gap are three significant areas that need to be addressed to make EEG-based control ready for daily use.

3 Neural control as part of the control architecture

3.1 Background

The control architecture for a steady motion is often divided into the high level and the low level (In some studies, the decision–making layer (or human intention perception), which determines the switching between different tasks (gaits), is defined as high level [1], but here we focus on a steady motion, which does not need this layer). The high-level controller determines the desired state of the device, which may consist of a combination of joint positions, velocities, and torques. In the low level control, these patterns will be translated to actuation commands (e.g., with PD for position control). The mechanical design and the actuation system of the device play an important role for designing an appropriate low level control. At the low-level control, torque, position, and impedance control can be implemented by using feedforward and/or feedback loops [1]. For example in Ref. [64], Zhang et al. compared nine different torque control approaches, including variations of classical feedback control, model-based control, adaptive control, and iterative learning. In order to investigate interactions between the desired torque (high level) and the tracking performance (low level), the authors tested each of these nine approaches in combination with four high-level controllers using time, joint angle, a neuromuscular model, or electromyographic measurements [64]. For all high-level control methods, they found that the combination of proportional control, damping injection and iterative learning could result in significantly lower error than other torque control approaches.

In this section we focus on high-level control, which can be divided into methods with and without neural control. We categorize common control approaches where the first three methods are without neural control as follows:

1. Predefined trajectory tracking-based approaches [65–67],
2. Predefined gait-pattern-based control [68,69],
3. Human-machine interaction model-based control. This approach considers gravity compensation and zero moment point balance criterion, and are mainly used for exoskeletons [70].
4. Adaptive oscillators-based control [basic concept used in central pattern generator (CPG)] [71]. (As one of neural control-based methods, this is described in more detail in the following).
5. Proportional myoelectrical control for single joint prostheses [54,72], (Section 2.3)
6. Neuromuscualr model-based approaches. Recent studies have shown that exoskeletons with a neuromuscular control comprising a dynamic model [73–75] can improve human walking assistance. We explain this method in more detail in the following.
7. Abstract model-based approaches. In this method, instead of complex models a simplified model is used. Virtual constraint control [76] and template-based control [77] are examples of abstract model-based approaches. The latter approach is described in detail in Section 4.

In some studies, hybrid strategies, including some of the aforementioned methods are used. For example, BLEEX adopts a force controller in swing phase and a position controller in stance phase [78]; or other studies on the combination of fuzzy control with other methods [79–81]. Since different aspects of high-level control strategies without neural control are covered in "Chapter 5, Disturbance observer applications in rehabilitation robotics: an overview" of this book, we concentrate on methods using neural control.

3.2 Neural control

One approach of realizing neural control is measuring and using human muscle activation signals. An overview about this topic is presented in Section 2.3. More details about myoelectric interfaces for control will be given in the next chapter. Here, we explain how a neural control concept can be implemented in control as a model-based approach to predict human movement and to provide movement in the assistive device that coordinates with the human motion. First, we briefly introduce the neural control and the neuromuscular models to explain human (animal) locomotion. For this, the CPG (central pattern generators) and the reflex-based control are described. As the main models to predict biological locomotion control, these are utilized to control assistive devices. At the end of this section, we present some examples of such applications.

3.2.1 Central pattern generators

During legged locomotion (e.g., walking and running), passive structures in the leg (e.g., Achilles tendon) store and release part of energy in each cycle. Active muscles should generate adequate force to compensate for energy losses needing appropriate control approaches. During the last few decades, the neuromuscular mechanisms, especially the central nervous system (CNS), are introduced as to explain how the activation pattern for skeletal muscles is produced in humans and animals. Such systems enable the biological locomotor systems to perform complex movements under different conditions.

The CNS hybrid automaton controlling structure consists of a low-level controller that generates a rhythmic repetitive patterns and a high-level controller that modulates the motor program [82]. How this system exactly provides activation for skeletal muscles during rhythmic motion (e.g., walking) is not completely known [83]. One phenomenon that has been identified is the ability to perform rhythmic activities in the spinal level, which are attributed to the CPGs. The CPG describes the low-level controlling role in CNS. That is responsible for generating muscle activation [84]. The CPGs are neural networks in the spinal cord (invertebrates), which have the ability to generate rhythmic outputs for leg extensor and flexor muscles regardless of the need for rhythmic inputs from the brain [85]. Different models of CPGs have been developed by researchers, which make them usable not only for controlling rhythmic movement in robots [86–90] but also for controlling the locomotion of biped robots [91–97]. The CPG neural control model is appropriate for controlling a limited number of muscles and joints [98]. For analyzing complex locomotion with higher number of contributing muscles, this model should be simplified. To do this, rather than activating muscles separately, a modular organization of motor control is suggested, called muscle synergies [99–104].

3.3 Neuromuscular model for control

Instead of using a model that has either CPG or target joint trajectories, Geyer et al. introduced a reflex-based neuromuscular model. In this model, the muscle reflexes can link the principles of leg and muscle-tendon dynamic systems and human motor control to generate stable locomotion [105–107]. The reflex-based motor control (using feedback signals from the muscle states) is an efficient and reliable alternative to central motor commands for generating cyclic locomotion. Geyer et al. proposed the idea that muscle reflexes can map sensory information, to the activation of leg muscles [108,105]. For example, they demonstrated that the stretch reflex amplifying muscle force during lengthening can control muscle-stiffness for generating stable hopping and running with a two segmented leg having one leg extensor muscle [108]. This idea of using positive force feedback was identified earlier in biological experiments [109]. The interplay between positive force feedback and the nonlinear muscle dynamics or (i.e., the interplay between the muscle reflexes with the environment) makes this muscle-reflex system self-adaptive.

To test if the method of controlling muscles based on autonomous reflex arcs can be generalized to human motor control, Geyer et al. developed a more detailed neuromuscular human model [105]. This way, the reflex pathways encode principles of leg dynamics and control. This model is based on the experimentally found compliant leg behavior and was implemented as a seven-segment system with 14 muscles. The muscle activities predicted by this detailed neuromuscular model were nearly the same as those reported from human walking data [105]. Furthermore, such a neuromuscular control can not only tolerate ground disturbances by thrusting or braking automatically, but it can also cope with terrain slopes without parameter interventions [107].

By using double leg dynamics as a fundamental model of swing leg dynamics, Geyer et al. developed the reflex-based control algorithm for robust swing leg placement to cope with large disturbances. Because of this enhancement, the generalization of the model to perform different locomotion tasks using spinal control model is achieved. Some instances are the steady and transitional locomotion behavior (walking and running), acceleration and deceleration, slope adaption and stair ambulation, and obstacle avoidance [106,110–112].

3.4 Application of neural control for gait assistance

As mentioned earlier, the key concept of CPG-based control is to view locomotion as a stable limit cycle behavior of the neural and mechanical systems modulated by simple control signals. To our knowledge, CPGs are not commonly used for controlling prosthetic legs. Instead, they are mostly applied to exoskeletons for gait assistance. CPG-based control [71] is widely used for subjects that can deliver periodic and stable locomotion, and are mostly applied to the hip joint actuation [71, 113–116].

In CPG-based control methods, the tuning and the correct parameterization of the controller for synchronization between different contributing elements to support various gait conditions is a challenge. On the other hand, a rhythmical sequence of locomotor-like contractions of a muscle could result from a sequence of spinal reflexes as found in flexor and extensor muscles of the hind limbs in spinally transected animals [117]. For example, the end of the swing phase generates a sensory signal to trigger the onset of the stance phase, or vice versa [117]. In Section 3.3, we explained that an alternative to CPG is the reflex-based control that can generate robust gaits without using predefined feedforward patterns [105,107].

As mentioned, the core idea of the reflex-based models is using feedback of muscle-fiber length, velocity, and force in combination with a muscle model to create the desired joint torque. Part of the reflex-based walking control was implemented in MITs powered prosthetic ankle [21]. The neuromuscular model exploiting an adaptive/versatile muscle reflex controller is examined by a knee–ankle prosthesis for a transfemoral amputee [118]. This model utilizes ankle plantar flexor, comprising Hill-type muscle model with positive force reflex for the prosthesis mimicking the behavior of the intact leg. To evaluate its performance, an initial simulation of an amputee walking with a powered prosthesis was constructed to assess its ability to produce robust locomotion over rough terrain without sensing the environment. Applying the neuromuscular model to a prosthetic leg, not only improved amputee gait stability, but also its robustness against terrain uncertainties over existing impedance control [118]. Recently, a reflex-based control using GRF as the feedback signal was implemented on the LOPES II exoskeleton [77]. This simplified method was inspired by a neuromuscular gait model utilizing positive force feedback [108,109]. In the next section, we describe this idea in more detail and how it can be used for control of a prosthetic ankle.

4 Template-based neural control

4.1 Why template-based control?

An impedance control, which uses simple mechanical elements (e.g., springdamper system) to predict human dynamic locomotion by exploiting a piecewise impedance function in each phase of the gait cycle, is one of the most common control schemes in powered prostheses [118,119]. This impedance function needs to adapt to the environment (e.g., terrains) in each phase of the gait cycle in order to tolerate undesired disturbances or to cope with changes in gait conditions (e.g., walking speed). For adapting to different gait conditions, scholars use a high-level hybrid controller by implementing a finite state machine. This method uses different states for each gait condition (e.g., ambulating stairs or level ground walking) and a sensory information (e.g., EMG signals) to detect user intention as a trigger to transit between these gait cycles [120]. In the previous section, we explained that another approach to deal with the adaptation problem is to use model-based bioinspired neuromuscular reflex control [118]. In this bioinspired method, the human musculoskeletal model and the neural (reflex-based) control are used to predict the human reaction for adapting to a different situation. Due to the complexity of the model-based neuromuscular model and the feedforward structure used in the impedance control, generating a universal bioinspired simple control method to assist healthy or impaired people in different gait condition will be hurdles. In order to reduce complexity, simplified biomechanical models (called "template" [121]) can be used for prediction and adaptation to different conditions (e.g., walking speed). Here, we present an extended version of template models, which uses neural control concept in addition to biomechanical templates. Hereafter, we use neuromechanical template models to emphasize a distinction to biomechanical templates. Some examples of such models are introduced in the force modulated compliant hip (FMCH) model [122] or the neuromuscular FMCH model [123]. In an application of neuromechanical template models for gait assistance [77,124], the advantage of impedance control is to simplify control to tuning the stiffness of a spring (and damper coefficient), The benefit of neuromuscular reflex control is its ability to adapt to different terrains [118]. With this approach, we can benefit from bioinspired control concept in a simplified (abstract) formulation. In the following, we present an example of such a method to describe how it can address adaptation to different gait conditions in a parsimonious control.

4.2 Neuromechanical-template-based control

Compared to biomechanical templates, neuromechanical template models use sensory information for control similar to the reflex-based control. Further, in neuromechanical template models, simple but adjustable mechanical elements (e.g., springs, damper, masses) are used to describe significant features of locomotion. Their properties (e.g. stiffness and damping) are tuned by using feedback signals such as length, velocity or force from the mechanical elements. Studies on normal and pathological gait demonstrate that humans use proprioceptive feedback signals in locomotion control [125]. For example, detecting load as a complex parameter, recorded by very different types of receptors, is crucial feedback for regulating joint movement and for shaping patterned motor output in biological gaits [126]. In addition, body loading is used as an input signal for adapting the locomotor patterns to external demands (gait condition) [127]. More importantly, Dietz et al. stated that "cyclical leg movements only in combination with loading of the legs lead to an appropriate leg muscle activation." This supports the idea of the significance of body load variation during the locomotor training in hemiplegic patients [128]. The leg force could continuously modulate the programmed pattern during locomotion according to information from peripheral receptors [129]. In addition to ability of measuring the load (body weight) in humans, high dependency of compensatory leg muscle activation was demonstrated experimentally [130]. Based on these insights, researchers try to use the GRF as a feedback signal for control of lower limb wearable robots. For example, the GRF was used to tune the hip spring stiffness in the FMCH model (force modulated hip compliance) to mimic human-like posture control [122]. This method was implemented on the LOPES II exoskeleton [77] to assist human walking. A recent study on this exoskeleton demonstrated the ability of the force feedback for reducing human metabolic cost and muscle activation during walking [131]. In a simulation study (using the neuromuscular model of Geyer [105]), the FMCH method was also employed for hip-knee assistance in an exosuit [124]. By introducing the concerted control concept, Sarmadi et al. extended the FMCH to coordinate balance and stance locomotor subfunctions (The general idea of locomotor subfunction is defining locomotion based on stance, swing, and balance subfunctions. In order to learn more about locomotor subfunction and applying it for gait assistance [132,133].) [134]. In the following section, we explain how the idea of using GRF as a reflex signal can be employed to control the ankle joint in a prosthetic foot. Such a force modulated compliant ankle (FMCA) concept was developed by the authors of this chapter.

4.3 FMCA concept

Load-compensating reflexes are highly flexible, and their gain can be adapted to the task or to the phase of the ongoing movement. As mentioned earlier, load plays a key role in shaping patterned motor output [126]. In the FMCA, a unique function can be identified to approximate the required ankle torque at different walking speeds. It was shown that ankle torque is dependent on walking speed [135] and the evolution of the ankle torque is highly correlated to the corresponding speed related GRF [136,137]. This dependency supports the assertion of using GRF as a useful sensory signal, which can be applied to the model to predict normal ankle joint torque without obtaining walking speed.

To estimate the ankle torque with FMCA, the stance phase of walking was divided into three states: the controlled plantarflexion (CP), the controlled dorsiflexion (CD), and the powered plantarflexion (PP) (Fig. 7.1). We consider an individual equation for each of the phases, to predict human ankle toque with the FMCA model. Further, for each walking speed, an individual variable impedance (defined by spring constant and damping coefficient) could be found to match human experimental ankle torque data. In the FMCA, tuning this variable impedance is formulated in a unified framework based on the measured vertical GRF. Hence, there is no need to change control parameters at different walking speeds. Fig. 7.2 shows the comparison between the measured ankle torque in human walking and the prediction using FMCA. Instead of using time-based patterns in

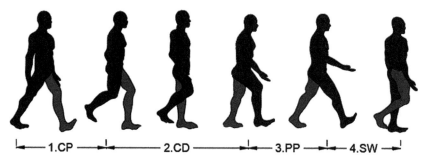

├──── 1.CP ────┼──────── 2.CD ────────┼── 3.PP ──┼── 4.SW ──┤

Figure 7.1 The human stride was divided into three states (subphases) for the stance phase beside the swing phase. The stance subphases: 1, controlled plantarflexion (CP); 2, controlled dorsiflexion (CD); and 3, powered plantarflexion (PP) are used to predict the human ankle torque with the FMCA model. CP starts with a neutral ankle position at heel strike and ends at foot flat when the ankle is most plantarflexed and the foot sole is completely placed on the ground. Weight acceptance and shock absorption occur in this state. CD starts at the end of the CP and ends at the beginning of the powered plantarflexion. During CD, the ankle reaches its maximum dorsiflexion. PP starts after CD and ends with the toe leaving the ground. During this phase, the positive ankle work is used for push-off. Swing (SW) starts at toe-off and ends with the heel strike. No GRF can be determined in this phase.

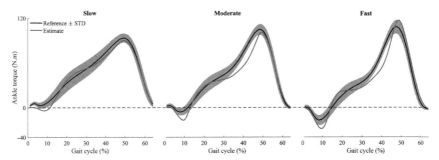

Figure 7.2 Mean and standard deviation of the human ankle torque based on experimental nonamputee data (*black*, [138]) and estimated ankle torque based on FMCA (*green*) (light gray in print version) for three different walking speeds (slow: 1.1 m/s, moderate: 1.55 m/s, and fast: 2.1 m/s).

the common trajectory tracking methods, here the GRF is employed as the phasing variable. Therefore, giving the FMCA-generated torque patterns to the prosthetic foot, releases dependency to time in trajectory-based methods. This force feedback includes other information (e.g., in case of perturbation occurrence), which is absent in methods using time-based trajectories.

4.3.1 Simulation

The simulation presented here uses the neuromuscular model including the prosthetic foot for one leg, developed by Thatte et al. [118]. Fig. 7.3 illustrates the finite state machine, defining the guard conditions for the

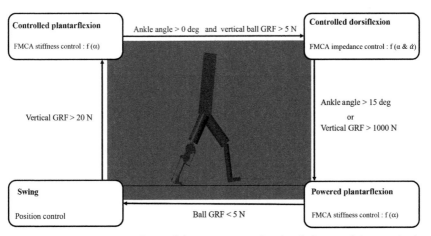

Figure 7.3 Finite state machine of the neuromuscular simulation model to apply the FMCA-based control scheme that modulates the human ankle torque during walking.

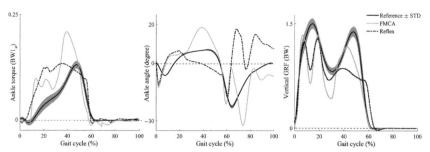

Figure 7.4 Comparison between control and performance of the reflex-based and FMCA approaches to control prosthetic foot, implemented in simulation model of [118]. The normalized ankle torque (to body weight BW and leg length in quiet standing *lst*), ankle angle, and vertical GRF normalized to BW, for reflex control (*blue*) (dark gray in print version) and template-based FMCA control (*green*) (light gray in print version) for one walking stride at 1.3 m/s (average normal walking speed) are compared with human experimental data (*black* [138]) at 1.55 m/s.

transition between states within the complete gait cycle, introduced in Ref. [118]. This figure depicts the switching conditions between CP, CD, PP, and swing phase, while the control parameters are optimized for each phase.

In the original simulation model, level transfemoral amputee walking was controlled by the reflex-based method [118]. We implemented the identified FMCA model from human walking on the neuromuscular simulation model, to control the ankle joint of a prosthetic foot. The remainder of the model (including the knee torque of the prosthesis controlled by the neuromuscular reflex method) is kept the same as the original one. To evaluate the FMCA concept, we demonstrate the ankle angle and exerted torque to the prosthetic foot next to the vertical GRF in Fig. 7.4. In these simulations, we found that the FMCA, with a simpler controller compared to the reflex-based method, can generate stable walking at normal speed. Furthermore, the FMCA is better than reflex-based method in mimicking experimental data. The ankle torque patterns are close to those of the human normal walking with small peak in the first half of and a big peak in the second half of the stance phase. Instead, with the reflex-based control, a plateaued region is observed in about 80% of the stance phase and the peak torque in the late stance to support push-off is missing. Comparing angle patterns confirms this finding, as the human-like (in unassisted walking) plantarflexion in the second half of the gait is missing in the reflex-based approach while it is pronounced in the FMCA. With the FMCA, the ankle dorsiflexion in the second half of stance phase is also more similar to normal walking patterns compared to the original model of Thatte et al. Interest-

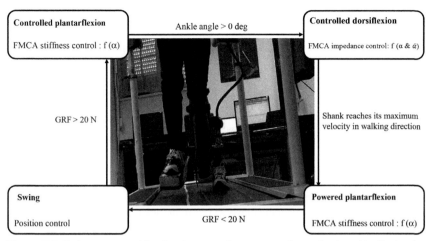

Figure 7.5 Finite state machine implemented on powered prosthetic ankle (SpringActive). The ankle angle and angular speed are shown by α and $\dot{\alpha}$.

ingly, the ankle angular movement in the swing phase with the FMCA controller is also more in line to natural walking than with the reflex-based control. As the swing phase control is the same in both models, this is mainly resulted from effects of the (late) stance phase control on the upcoming leg swinging. In both simulation models, the GRF have three-hump instead of double-hump in human walking. Basically, the first two humps support the loading phase and approximate the first peak in normal walking GRF pattern. However, the third hump found in FMCA controlled is more consistent with the second hump in normal walking than that of the reflex-based control. Again, lack of human-like push-off power in the reflex-based method results in diminished peak in the unloading phase, which is resolved in the FMCA approach. In summary, the simulation results support not only the applicability of the proposed neuromechanical template method, but they also result in more natural gaits compared to the neuromuscular controller while the formulation is much simpler.

4.3.2 Human experiments

After analyzing the FMCA method with the neuromuscular simulation model, this method was implemented on a powered prosthetic ankle (Ruggedized Odyssey Ankle, by Springactive). The main goal was to evaluate the ability of FMCA to generate human-like control patterns in the ankle joint (e.g., torque and power), especially for push-off. Vertical GRF was measured by an instrumented treadmill. As shown in Fig. 7.5, the experiment was conducted by a nonamputee with the help of an orthotic bypass adapter.

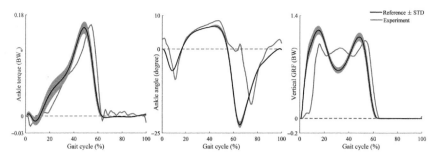

Figure 7.6 The normalized ankle torque (to body weight BW and leg length in quiet standing *lst*), ankle angle, and vertical GRF normalized to BW in walking. Mean and standard deviation of human walking experimental data (reference in *black*) at 1.55 m/s from [138]) is compared to walking data with the powered prosthetic ankle (experiment in *green*) (light gray in print version) at 1.3 m/s, controlled with FMCA, both on an instrumented treadmill.

The contralateral leg length was increased by additional foam, mounted below the shoe.

Similar to the simulation model, the stance phase was divided into three aforementioned states (subphases) and the swing phase, and a finite state machine was used to switch between them. Fig. 7.5 depicts the transition between different states. Shank angular velocity, which is sensed by gyro sensor, replaced the GRF as a guard condition for the transition from CD to PP. Although, GRF measured by the instrumented treadmill could be utilized, we employed the angular velocity data for detecting PP subphase to provide a fair comparison with the trajectory-based methods (results are not presented here). Torque generation based on the FMCA model is used in the whole stance phase. This includes different GRF-based torque formulations in the three subphases of stance phase. Energy injection is mainly performed in the PP phase, which is the most important period for assisting the impaired leg (here, the leg with bypassed ankle).

We assumed that the FMCA template model, which is a simplified abstraction of a complex neuromuscular model, can provide a control solution to mimic human dynamical behavior in the ankle joint. In the pilot experiments at 1.3 m/s, equal to 75% of the preferred transition speed (PTS) of the subject, the outcomes supports this hypothesis. The vertical GRF, ankle angle, and ankle torque were similar to those reported for nonamputee level ground walking data at 75% PTS (equal to 1.55 m/s for the average of 21 subjects [138]), as shown in Fig. 7.6. In spite of all discrepancies between normal walking and assisted walking with a different configuration in the leg (e.g., elevated foot, bypassed ankle, outward shifting of the center of pressure by the bypass system shown in Fig. 7.5), the results are comparable.

These results support the applicability of the neuromechanical template model-based control method for gait assistance in the prostheses.

5 SUMMARY

The chapter presented three different areas of neural control including neural input signals, neural control as part of the control architecture, and template-based neural models. While EMG is the most common neural signal investigated in lower limb prosthetic research, there is no commercial powered lower limb prostheses using it as of yet. Reasons for this are manifold, for example, challenges in the acquisition and the processing of the data that can potentially result in unreliable data. Next to neural input signals, neural control as a part of the control architecture was introduced. Such approaches have already proven their applicability for daily use in a powered prosthetic ankle. Simplified template-based neural concepts were introduced by the authors. First study results seem promising for the use of such concepts for real world applications.

References

[1] M.R. Tucker, J. Olivier, A. Pagel, H. Bleuler, M. Bouri, O. Lambercy, J. del, R. Millán, R. Riener, H. Vallery, R. Gassert, Control strategies for active lower extremity prosthetics and orthotics: a review, J. Neuroeng. Rehabilit. 12 (1) (2015) 1.

[2] R.R. Torrealba, G. Fernández-López, J.C. Grieco, Towards the development of knee prostheses: review of current researches, Kybernetes 37 (9/10) (2008) 1561–1576.

[3] R. Jimenez-Fabian, O. Verlinden, Review of control algorithms for robotic ankle systems in lower-limb orthoses, prostheses, and exoskeletons, Med. Eng. Physics 34 (4) (2012) 397–408.

[4] M. Grimmer, A. Seyfarth, Mimicking human-like leg function in prosthetic limbs, in: Neuro-Robotics, Dordrecht, The Netherlands, Springer, 2014, pp. 105–155.

[5] P. Cherelle, G. Mathijssen, Q. Wang, B. Vanderborght, D. Lefeber, Advances in propulsive bionic feet and their actuation principles, Adv. Mech. Eng. 6 (2014) 984046.

[6] M. Windrich, M. Grimmer, O. Christ, S. Rinderknecht, P. Beckerle, Active lower limb prosthetics: a systematic review of design issues and solutions, Biomed. Eng. Online 15 (3) (2016) 140.

[7] C. Ferreira, L.P. Reis, C.P. Santos, Review of control strategies for lower limb prostheses, in: Robot 2015: Second Iberian Robotics Conference, 2016, Springer, pp. 209–220.

[8] D.S. Pieringer, M. Grimmer, M.F. Russold, R. Riener, Review of the actuators of active knee prostheses and their target design outputs for activities of daily living, in: International Conference on Rehabilitation Robotics (ICORR), IEEE, 2017, pp. 1246–1253.

[9] T.G. Sugar, J.A. Ward, M. Grimmer, Ankle prosthetics and orthotics: trends from passive to active systems, in: The Encyclopedia of Medical Robotics, 2018, World Scientific, pp. 111–134.

[10] C.M. Lara-Barrios, A. Blanco-Ortega, C.H. Guzmán-Valdivia, K.D. Bustamante Valles, Literature review and current trends on transfemoral powered prosthetics, Adv. Robot. 32 (2) (2018) 51–62.

[11] K. Zhang, C.W. de Silva, C. Fu, Sensor fusion for predictive control of human-prosthesis-environment dynamics in assistive walking: a survey, CoRR, abs/1903.07674, 2019, arXiv preprint arXiv:1903.07674.

[12] J.K. Hitt, R. Bellman, M. Holgate, T.G. Sugar, K.W. Hollander, The sparky (spring ankle with regenerative kinetics) project: design and analysis of a robotic transtibial prosthesis with regenerative kinetics, in: ASME 2007 International Design Engineering Technical Conferences and Computers and Information in Engineering Conference, 2007, American Society of Mechanical Engineers, pp. 1587–1596.

[13] D. Quintero, D.J. Villarreal, R.D. Gregg, Preliminary experiments with a unified controller for a powered knee-ankle prosthetic leg across walking speeds, in: IEEE/RSJ International Conference on Intelligent Robots and Systems (IROS), IEEE, 2016, pp. 5427–5433.

[14] H.M. Herr, A.M. Grabowski, Bionic ankle–foot prosthesis normalizes walking gait for persons with leg amputation, Proc. Royal Soc. B Biol. Sci. 279 (1728) (2011) 457–464.

[15] A. H. Shultz, J.E. Mitchell, D. Truex, B.E. Lawson, M. Goldfarb, Preliminary evaluation of a walking controller for a powered ankle prosthesis, in: IEEE International Conference on Robotics and Automation, IEEE, 2013, pp. 4838–4843.

[16] M. Grimmer, M. Holgate, J. Ward, A. Seyfarth, Feasibility study of transtibial amputee walking using a powered prosthetic foot, in: International Conference on Rehabilitation Robotics (ICORR), IEEE, 2017, pp. 1118–1123.

[17] F. Sup, H.A. Varol, J. Mitchell, T. Withrow, M. Goldfarb, Design and control of an active electrical knee and ankle prosthesis, in: 2nd IEEE RAS & EMBS International Conference on Biomedical Robotics and Biomechatronics, IEEE, 2008, pp. 523–528.

[18] D. Grimes, W. Flowers, M. Donath, Feasibility of an active control scheme for above knee prostheses, J. Biomec. Eng. 99 (4) (1977) 215–221.

[19] J. Zhu, Q. Wang, L. Wang, Pantoe 1: biomechanical design of powered ankle-foot prosthesis with compliant joints and segmented foot, in: IEEE/ASME International Conference on Advanced Intelligent Mechatronics, IEEE, 2010, pp. 31–36.

[20] P.B. Shull, W. Jirattigalachote, M.A. Hunt, M.R. Cutkosky, S.L. Delp, Quantified self and human movement: a review on the clinical impact of wearable sensing and feedback for gait analysis and intervention, Gait Post. 40 (1) (2014) 11–19.

[21] M.F. Eilenberg, H. Geyer, H. Herr, Control of a powered ankle–foot prosthesis based on a neuromuscular model, TNSRE 18 (2) (2010) 164–173.

[22] M. Grimmer, M. Holgate, R. Holgate, A. Boehler, J. Ward, K. Hollander, T. Sugar, A. Seyfarth, A powered prosthetic ankle joint for walking and running, Biomed. Eng. Online 15 (3) (2016) 141.

[23] P. Cherelle, A. Matthys, V. Grosu, B. Brackx, M. Van Damme, B. Vanderborght, D. Lefeber, The amp-foot 2.0: a powered transtibial prosthesis that mimics intact ankle behavior, in: 9th National Congress on Theoretical and Applied Mechanics, 2012, Brussels, Citeseer, pp. 1–5.

[24] R. Versluys, G. Lenaerts, M. Van Damme, I. Jonkers, A. Desomer, B. Vanderborght, L. Peeraer, G. Van der Perre, D. Lefeber, Successful preliminary walking experiments on a transtibial amputee fitted with a powered prosthesis, Prosthet. Orthot. Int. 33 (4) (2009) 368–377.

[25] B.G. Lambrecht, H. Kazerooni, Design of a semi-active knee prosthesis, in: IEEE International Conference on Robotics and Automation, IEEE, 2009, pp. 639–645.

[26] F. Zhang, Z. Fang, M. Liu, H. Huang, Preliminary design of a terrain recognition system, in: Conference proceedings:... Annual International Conference of the IEEE Engineering in Medicine and Biology Society. IEEE Engineering in Medicine and Biology Society. Annual Conference, vol. 2011, NIH Public Access, 2011, pp. 5452–5455.

[27] M. Liu, D. Wang, H.H. Huang, Development of an environment-aware locomotion mode recognition system for powered lower limb prostheses, TNSRE 24 (4) (2015) 434–443.

[28] G. G. Scandaroli, G.A. Borges, J.Y. Ishihara, M.H. Terra, A.F. da Rocha, F.A. de Oliveira Nascimento, Estimation of foot orientation with respect to ground for an above knee robotic prosthesis, in: IEEE/RSJ International Conference on Intelligent Robots and Systems, IEEE, 2009, pp. 1112–1117.

[29] N. Krausz, T. Lenzi, L. Hargrove, Depth sensing for improved control of lower limb prostheses, IEEE Transact. Bio-Med. Eng. 62 (11) (2015) 2576–2587.

[30] J. P. Diaz, R.L. da Silva, B. Zhong, H.H. Huang, E. Lobaton, Visual terrain identification and surface inclination estimation for improving human locomotion with a lower-limb prosthetic, in: 40th Annual International Conference of the IEEE Engineering in Medicine and Biology Society (EMBC), IEEE, 2018, pp. 1817–1820.

[31] J.-Y. Guo, Y.-P. Zheng, H.-B. Xie, X. Chen, Continuous monitoring of electromy-ography (EMG), mechanomyography (MMG), sonomyography (SMG) and torque output during ramp and step isometric contractions, Med. Eng. Phys. 32 (9) (2010) 1032–1042.

[32] C.M. Smith, T.J. Housh, E.C. Hill, J.L. Keller, G.O. Johnson, R.J. Schmidt, A biosignal analysis for reducing prosthetic control durations: a proposed method using electro-myographic and mechanomyographic control theory, JMNI 19 (2) (2019) 142–149.

[33] J. Ribeiro, F. Mota, T. Cavalcante, I. Nogueira, V. Gondim, V. Albuquerque, A. Alexandria, Analysis of man-machine interfaces in upper-limb prosthesis: a review, Robotics 8 (1) (2019) 16.

[34] N. Akhlaghi, C.A. Baker, M. Lahlou, H. Zafar, K.G. Murthy, H.S. Rangwala, J. Kosecka, W.M. Joiner, J.J. Pancrazio, S. Sikdar, Real-time classification of hand motions using ultrasound imaging of forearm muscles, IEEE Transact. Biomed. Eng. 63 (8) (2015) 1687–1698.

[35] D. Sierra González, C. Castellini, A realistic implementation of ultrasound imaging as a human-machine interface for upper-limb amputees, Front. Neurorobot. 7 (2013) 17.

[36] J. Zhang, P. Fiers, K.A. Witte, R.W. Jackson, K.L. Poggensee, C.G. Atkeson, S.H. Collins, Human-in-the-loop optimization of exoskeleton assistance during walking, Science 356 (6344) (2017) 1280–1284.

[37] Y. Ding, M. Kim, S. Kuindersma, C.J. Walsh, Human-in-the-loop optimization of hip assistance with a soft exosuit during walking, Sci. Robot. 3 (15) (2018) eaar5438.

[38] K.A. Ingraham, D.P. Ferris, C.D. Remy, Evaluating physiological signal salience for es-timating metabolic energy cost from wearable sensors, J. Appl. Physiol 126 (3) (2019) 717–729.

[39] X. Navarro, T.B. Krueger, N. Lago, S. Micera, T. Stieglitz, P. Dario, A critical review of interfaces with the peripheral nervous system for the control of neuroprostheses and hybrid bionic systems, J. Perip. Nerv. Sys. 10 (3) (2005) 229–258.

[40] A. Fougner, Ø. Stavdahl, P.J. Kyberd, Y.G. Losier, P.A. Parker, Control of upper limb prostheses: terminology and proportional myoelectric control—a review, TNSRE 20 (5) (2012) 663–677.

[41] G. Horn, Electro-control: am emg-controlled a/k prosthesis, Med. Biol. Eng. 10 (1) (1972) 61–73.

[42] W. Dyck, S. Onyshko, D. Hobson, D. Winter, A. Quanbury, A voluntarily controlled electrohydraulic above-knee prosthesis, Bull. Prosthet. Res. Spring (1975) 169.

[43] K.H. Ha, H.A. Varol, M. Goldfarb, Volitional control of a prosthetic knee using surface electromyography, IEEE Transact. Biomed. Eng. 58 (1) (2011) 144–151.

[44] L. Hargrove, A.M. Simon, S.B. Finucane, R.D. Lipschutz, Myoelectric control of a pow-ered transfemoral prosthesis during non-weight-bearing activities, in: Proceedings of the 2011 MyoElectric Controls/Powered Prosthetics Symposium, Fredericton, Myo-electric Symposium, 2011, pp. 1–41.

[45] H. Huang, T.A. Kuiken, R.D. Lipschutz, et al. A strategy for identifying locomotion modes using surface electromyography, IEEE Transact. on Biomedical Engineering 56 (1) (2009) 65–73.

[46] S.-K. Wu, G. Waycaster, X. Shen, Electromyography-based control of active above-knee prostheses, Control Eng. Pract. 19 (8) (2011) 875–882.

[47] J. Wang, O.A. Kannape, H.M. Herr, Proportional emg control of ankle plantar flexion in a powered transtibial prosthesis, in: IEEE 13th International Conference on Rehabilitation Robotics (ICORR), IEEE, 2013, pp. 1–5.

[48] X. Guo, L. Chen, Y. Zhang, P. Yang, L.-Q. Zhang, A study on control mechanism of above knee robotic prosthesis based on CPG model, in: IEEE International Conference on Robotics and Biomimetics, ROBIO, 2010, pp. 283–287.

[49] C.D. Hoover, G.D. Fulk, K.B. Fite, Stair ascent with a powered transfemoral prosthesis under direct myoelectric control, IEEE/ASME Transact. Mechat. 18 (3) (2012) 1191–1200.

[50] M. R. Islam, A.M. Haque, S. Amin, K. Rabbani, Design and development of an emg driven microcontroller based prosthetic leg, IOMP 4 (1), 107–114.

[51] S. K. Au, P. Bonato, H. Herr, An emg-position controlled system for an active ankle-foot prosthesis: an initial experimental study, in: 9th International Conference on Rehabilitation Robotics, IEEE, 2005, pp. 375–379.

[52] A.L. Delis, J.L. Carvalho, A.F. Da Rocha, F.A. Nascimento, G.A. Borges, Myoelectric knee angle estimation algorithms for control of active transfemoral leg prostheses, in: Self Organizing Maps-Applications and Novel Algorithm Design, IntechOpen, 2011, pp. 401–424.

[53] C. Fleischer, C. Reinicke, G. Hommel, Predicting the intended motion with emg signals for an exoskeleton orthosis controller, in: IEEE/RSJ International Conference on Intelligent Robots and Systems, IEEE, 2005, pp. 2029–2034.

[54] C. Fleischer, G. Hommel, A human–exoskeleton interface utilizing electromyography, IEEE Transact. Robot. 24 (4) (2008) 872–882.

[55] C.D. Hoover, G.D. Fulk, K.B. Fite, The design and initial experimental validation of an active myoelectric transfemoral prosthesis, J. Med. Dev. 6 (1) (2012) 011005.

[56] M. Donath, Proportional emg control for above knee pros-theses, PhD thesis. Massachusetts Institute of Technology, 1974.

[57] S. Micera, A.M. Sabatini, P. Dario, B. Rossi, A hybrid approach to emg pattern analysis for classification of arm movements using statistical and fuzzy techniques, Med. Eng. Phys. 21 (5) (1999) 303–311.

[58] M. Zecca, S. Micera, M.C. Carrozza, P. Dario, Control of multifunctional prosthetic hands by processing the electromyographic signal, Crit. Rev. Biomed. Eng. 30 (4–6) (2002) 459–485.

[59] R.E. Alcaide-Aguirre, D.C. Morgenroth, D.P. Ferris, Motor control and learning with lower-limb myoelectric control in amputees, J. Rehabil. Res. Dev. 50 (5) (2013) 687–698.

[60] H. Huang, F. Zhang, L.J. Hargrove, Z. Dou, D.R. Rogers, K.B. Engle hart, Continuous locomotion-mode identification for prosthetic legs based on neuromuscular–mechanical fusion, IEEE Transac. Biomed. Eng. 58 (10) (2011) 2867–2875.

[61] M. AL-Quraishi, I. Elamvazuthi, S. Daud, S. Parasuraman, A. Borboni, EEG-based control for upper and lower limb exoskeletons and prostheses: a systematic review, Sensors 18 (10) (2018) 3342.

[62] Y. He, D. Eguren, J.M. Azorín, R.G. Grossman, T.P. Luu, J.L. Contreras-Vidal, Brain–machine interfaces for controlling lower-limb powered robotic systems, J. Neural Eng. 15 (2018) 15p.

[63] D.P. Murphy, O. Bai, A.S. Gorgey, J. Fox, W.T. Lovegreen, B.W. Burkhardt, R. Atri, J.S. Marquez, Q. Li, D.-Y. Fei, Electroencephalogrambased brain–computer interface and lower-limb prosthesis control: a case study, Front. Neurol. 8 (2017) 696.

[64] J. Zhang, C.C. Cheah, S.H. Collins, Torque control in legged locomotion, in: M.A., Sharbafi, A., Seyfarth (Eds.), Bioinspired Legged Locomotion, Elsevier, 2017, Ch. 5, pp. 347–400.

[65] D. Sanz-Merodio, M. Cestari, J.C. Arevalo, E. Garcia, Control motion approach of a lower limb orthosis to reduce energy consumption, IJARS 9 (6) (2012) 232.

[66] S. Wang, L. Wang, C. Meijneke, E. van Asseldonk, T. Hoellinger, G. Cheron, Y. Ivanenko, V.L. Scaleia, F. Sylos-Labini, M. Molinari, F. Tamburella, I. Pisotta, F. Thorsteinsson, M. Ilzkovitz, J. Gancet, Y. Nevatia, R. Hauffe, F. Zanow, H. van der Kooij, Design and control of the mindwalker exoskeleton, TNSRE 23 (2) (2015) 277–286, doi: 10.1109/TNSRE. 2014.2365697.

[67] K. A. Strausser, H. Kazerooni, The development and testing of a human machine interface for a mobile medical exoskeleton, in: IEEE/RSJ International Conference on Intelligent Robots and Systems (IROS), IEEE, 2011, pp. 4911–4916.

[68] A.T. Asbeck, R.J. Dyer, A.F. Larusson, C.J. Walsh, Biologicallyinspired soft exosuit, in: IEEE international conference on Rehabilitation robotics (ICORR), IEEE, 2013, pp. 1–8.

[69] C.J. Walsh, K. Pasch, H. Herr, An autonomous, underactuated exoskeleton for load-carrying augmentation, in: IEEE/RSJ International Conference on Intelligent Robots and Systems, IEEE, 2006, pp. 1410–1415.

[70] T. Yan, M. Cempini, C.M. Oddo, N. Vitiello, Review of assistive strategies in powered lower-limb orthoses and exoskeletons, Robot. Auton. Sys. 64 (2015) 120–136.

[71] R. Ronsse, T. Lenzi, N. Vitiello, B. Koopman, E. van Asseldonk, S.M.M. De Rossi, J. van den Kieboom, H. van der Kooij, M.C. Carrozza, A.J. Ijspeert, Oscillator-based assistance of cyclical movements: model-based and model-free approaches, Med. Biol. Eng. Comput. 49 (10) (2011) 1173.

[72] P.-C. Kao, C.L. Lewis, D.P. Ferris, Invariant ankle moment patterns when walking with and without a robotic ankle exoskeleton, J. Biomech. 43 (2) (2010) 203–209.

[73] A.R. Wu, F. Dzeladini, T.J.H. Brug, F. Tamburella, N.L. Tagliamonte, E.H.F. van Asseldonk, H. van der Kooij, A.J. Ijspeert, An adaptive neuromuscular controller for assistive lower-limb exoskeletons: a preliminary study on subjects with spinal cord injury, Front. Neurorobot. 11 (2017) 30, doi: 10.3389/fnbot.2017.00030.

[74] F. Dzeladini, A.R. Wu, D. Renjewski, A. Arami, E. Burdet, E. van Asseldonk, H. van der Kooij, A.J. Ijspeert, Effects of a neuromuscular controller on a powered ankle exoskeleton during human walking, in: 6th IEEE International Conference on Biomedical Robotics and Biomechatronics (BioRob), 2016, pp. 617–622. doi:10.1109/BIOROB. 2016.7523694.

[75] V. Ruiz Garate, A. Parri, T. Yan, M. Munih, R. Molino Lova, N. Vitiello, R. Ronsse, Walking assistance using artificial primitives: a novel bioinspired framework using motor primitives for locomotion assistance through a wearable cooperative exoskeleton, IEEE Robot. Automat. Mag. 23 (1) (2016) 83–95, doi: 10.1109/MRA. 2015.2510778.

[76] R.D. Gregg, J.W. Sensinger, Towards biomimetic virtual constraint control of a powered prosthetic leg, IEEE Transact. Control Syst. Technol. 22 (1) (2013) 246–254.

[77] G. Zhao, M. Sharbafi, M. Vlutters, E. Van Asseldonk, A. Seyfarth, Template model inspired leg force feedback based control can assist human walking, in: International Conference on Rehabilitation Robotics (ICORR), IEEE, 2017, pp. 473–478.

[78] H. Kazerooni, R. Steger, L. Huang, Hybrid control of the berkeley lower extremity exoskeleton (bleex), Int. J. Robot. Res. 25 (5–6) (2006) 561–573.

[79] T.-S. Wu, M. Karkoub, H.-S. Chen, W.-S. Yu, M.-G. Her, Robust tracking observer-based adaptive fuzzy control design for uncertain nonlinear mimo systems with time delayed states, Informat. Sci. 290 (2015) 86–105.

[80] D. Wang, M. Liu, F. Zhang, H. Huang, Design of an expert system to automatically calibrate impedance control for powered knee prostheses, in: IEEE 13th International Conference on Rehabilitation Robotics (ICORR), IEEE, 2013, pp. 1–5.

[81] T. Lenzi, M.C. Carrozza, S.K. Agrawal, Powered hip exoskeletons can reduce the user's hip and ankle muscle activations during walking, TNSRE 21 (6) (2013) 938–948.

[82] I.A. Rybak, D.G. Ivashko, B.I. Prilutsky, M.A. Lewis, J.K. Chapin, Modeling neural control of locomotion: integration of reflex circuits with CPG, in: International Conference on Artificial Neural Networks, Springer, 2002, pp. 99–104.

[83] V. Dietz, J. Quintern, G. Boos, W. Berger, Obstruction of the swing phase during gait: phase-dependent bilateral leg muscle coordination, Brain Res. 384 (1) (1986) 166–169.

[84] S.A. Haghpanah, F. Farahmand, H. Zohoor, Modular neuromuscular control of human locomotion by central pattern generator, J. Biomech. 53 (2017) 154–162.

[85] A.J. Ijspeert, Central pattern generators for locomotion control in animals and robots: a review, Neural Net. 21 (4) (2008) 642–653.

[86] C. Liu, Q. Chen, D. Wang, Cpg-inspired workspace trajectory generation and adaptive locomotion control for quadruped robots, IEEE Trans. Syst. Man. Cybern. B. Cybern. 41 (3) (2011) 867–880.

[87] Y. Fukuoka, H. Kimura, A.H. Cohen, Adaptive dynamic walking of a quadruped robot on irregular terrain based on biological concepts, Int. J. Robot. Res. 22 (3–4) (2003) 187–202.

[88] J.-K. Ryu, N.Y. Chong, B.J. You, H.I. Christensen, Locomotion of snake-like robots using adaptive neural oscillators, Int. Ser. Robot. 3 (1) (2010) 1.

[89] H. Kimura, Y. Fukuoka, A.H. Cohen, Adaptive dynamic walking of a quadruped robot on natural ground based on biological concepts, Int. J. Robot. Res. 26 (5) (2007) 475–490.

[90] G.L. Liu, M.K. Habib, K. Watanabe, K. Izumi, Central pattern generators based on matsuoka oscillators for the locomotion of biped robots, Artif. Life Robot. 12 (1–2) (2008) 264–269.

[91] A.J. Ijspeert, J. Kodjabachian, Evolution and development of a central pattern generator for the swimming of a lamprey, Artif. Life 5 (3) (1999) 247–269.

[92] C. Maufroy, H. Kimura, K. Takase, Towards a general neural controller for quadrupedal locomotion, Neural Net. 21 (4) (2008) 667–681.

[93] D. Sussillo, L.F. Abbott, Generating coherent patterns of activity from chaotic neural networks, Neuron 63 (4) (2009) 544–557.

[94] G. Taga, Y. Yamaguchi, H. Shimizu, Self-organized control of bipedal locomotion by neural oscillators in unpredictable environment, Biol. Cybernet. 65 (3) (1991) 147–159.

[95] J.S. Bay, H. Hemami, Modeling of a neural pattern generator with coupled nonlinear oscillators, IEEE Transact. Biomed. Eng. 34 (4) (1987) 297–306.

[96] N.G. Hatsopoulos, Coupling the neural and physical dynamics in rhythmic movements, Neural Comput. 8 (3) (1996) 567–581.

[97] C.P. Santos, V. Matos, CPG modulation for navigation and omnidirectional quadruped locomotion, Robot. Autonom. Sys. 60 (6) (2012) 912–927.

[98] S.N. Markin, A.N. Klishko, N.A. Shevtsova, M.A. Lemay, B.I. Prilutsky, A. Rybak, Afferent control of locomotor CPG: insights from a simple neuromechanical model, Ann. N.Y. Acad. Sci. 1198 (1) (2010) 21–34.

[99] J.L. Allen, R.R. Neptune, Three-dimensional modular control of human walking, J. Biomech. 45 (12) (2012) 2157–2163.

[100] C.P. McGowan, R.R. Neptune, D.J. Clark, S.A. Kautz, Modular control of human walking: adaptations to altered mechanical demands, J. Biomech. 43 (3) (2010) 412–419.

[101] R.R. Neptune, D.J. Clark, S.A. Kautz, Modular control of human walking: a simulation study, J. Biomech. 42 (9) (2009) 1282–1287.

[102] Y.P. Ivanenko, G. Cappellini, N. Dominici, R.E. Poppele, F. Lacquaniti, Coordination of locomotion with voluntary movements in humans, J. Neurosci. 25 (31) (2005) 7238–7253.

[103] G. Cappellini, Y.P. Ivanenko, R.E. Poppele, F. Lacquaniti, Motor patterns in human walking and running, J. Neurophysiol. 95 (6) (2006) 3426–3437.

[104] M.C. Tresch, V.C. Cheung, A. d'Avella, Matrix factorization algorithms for the identification of muscle synergies: evaluation on simulated and experimental data sets, J. Neurophysiol. 95 (4) (2006) 2199–2212.

[105] H. Geyer, H. Herr, A muscle-reflex model that encodes principles of legged mechanics produces human walking dynamics and muscle activities, TNSRE 18 (3) (2010) 263–273.

[106] S. Song, H. Geyer, Generalization of a muscle-reflex control model to 3d walking, in: 35th Annual International Conference of the IEEE Engineering in Medicine and Biology Society (EMBC), IEEE, 2013, pp. 7463–7466.

[107] S. Song, H. Geyer, A neural circuitry that emphasizes spinal feedback generates diverse behaviours of human locomotion, J. Physiol. 593 (16) (2015) 3493–3511.

[108] H. Geyer, A. Seyfarth, R. Blickhan, Positive force feedback in bouncing gaits?, Proc. Royal Soc. London Series B Biol. Sci. 270 (1529) (2003) 2173–2183.

[109] A. Prochazka, D. Gillard, D.J. Bennett, Positive force feedback control of muscles, J. Neurophysiol. 77 (6) (1997) 3226–3236.

[110] R. Desai, H. Geyer, Robust swing leg placement under large disturbances, in: IEEE International Conference on Robotics and Biomimetics (ROBIO), IEEE, 2012, pp. 265–270.

[111] R. Desai, H. Geyer, Muscle-reflex control of robust swing leg placement, in: IEEE International Conference on Robotics and Automation, IEEE, 2013, pp. 2169–2174.

[112] S. Song, R. Desai, H. Geyer, Integration of an adaptive swing control into a neuromuscular human walking model, in: 35th Annual International Conference of the IEEE Engineering in Medicine and Biology Society (EMBC), IEEE, 2013, pp. 4915–4918.

[113] F. Giovacchini, F. Vannetti, M. Fantozzi, M. Cempini, M. Cortese, A. Parri, T. Yan, D. Lefeber, N. Vitiello, A light-weight active orthosis for hip movement assistance, Robot. Auton. Syst. 73 (2015) 123–134.

[114] N. L. Tagliamonte, F. Sergi, G. Carpino, D. Accoto, E. Guglielmelli, Human-robot interaction tests on a novel robot for gait assistance, in: IEEE International Conference on Rehabilitation Robotics (ICORR), IEEE, 2013, pp. 1–6.

[115] T. Matsubara, A. Uchikata, J. Morimoto, Full-body exoskeleton robot control for walking assistance by style-phase adaptive pattern generation, in: IEEE/RSJ International Conference on Intelligent Robots and Systems (IROS), IEEE, 2012, pp. 3914–3920.

[116] X. Zhang, M. Hashimoto, Synchronization based control for walking assist suit-evaluation on synchronization and assist effect. Key Eng. Mat. 464, 115–118.

[117] A. Freusbcrg, Reflexbewegungen beim hunde, Pflügers Archiv Eur. J. Physiol. 9 (1) (1874) 358–391.

[118] N. Thatte, H. Geyer, Toward balance recovery with leg prostheses using neuromuscular model control, IEEE Transact. Biomed. Eng. 63 (5) (2016) 904–913.

[119] F. Sup, H.A. Varol, J. Mitchell, T.J. Withrow, M. Goldfarb, Preliminary evaluations of a self-contained anthropomorphic transfemoral prosthesis, IEEE/ASME T. Mech. 14 (6) (2009) 667–676.

[120] S. Au, M. Berniker, H. Herr, Powered ankle-foot prosthesis to assist levelground and stair-descent gaits, Neural Net. 21 (4) (2008) 654–666.

[121] R.J. Full, D. Koditschek, Templates and anchors: neuromechanical hypotheses of legged locomotion on land, J. Exp. Biol. 22 (1999) 3325–3332.

[122] M.A. Sharbafi, A. Seyfarth, Fmch: a new model for human-like postural control in walking, in: IEEE/RSJ International Conference on Intelligent Robots and Systems (IROS), IEEE, 2015, pp. 5742–5747.

[123] A. Davoodi, O. Mohseni, A. Seyfarth, M.A. Sharbafi, From template to anchors: transfer of virtual pendulum posture control balance template to adaptive neuromuscular gait model increases walking stability, Royal Soc. Open Sci. 6 (3) (2019) 181911.

[124] M.A. Sharbafi, H. Barazesh, M. Iranikhah, A. Seyfarth, Leg force control through biarticular muscles for human walking assistance, Front. Neurorobot 12, 39.

[125] V. Dietz, K. Leenders, G. Colombo, Leg muscle activation during gait in parkinson's disease: influence of body unloading, Electroencephalogr. Clin. Neurophysiol. 105 (5) (1997) 400–405.

[126] J. Duysens, F. Clarac, H. Cruse, Load-regulating mechanisms in gait and posture: comparative aspects, Physiol. Rev. 80 (1) (2000) 83–133.

[127] V. Dietz, R. Müller, G. Colombo, Locomotor activity in spinal man: significance of afferent input from joint and load receptors, Brain 125 (12) (2002) 2626–2634.

[128] S. Hesse, B. Helm, J. Krajnik, M. Gregoric, K.H. Mauritz, Treadmill training with partial body weight support: influence of body weight release on the gait of hemiparetic patients, J. Neurol. Rehabilit. 11 (1) (1997) 15–20.

[129] V. Dietz, J. Duysens, Significance of load receptor input during locomotion: a review, Gait Post. 11 (2) (2000) 102–110.

[130] V. Dietz, G. Horstmann, M. Trippel, A. Gollhofer, Human postural reflexes and gravityan under water simulation, Neurosci. Lett. 106 (3) (1989) 350–355.

[131] G. Zhao, M.A. Sharbafi, M. Vlutters, E. Van Asseldonk, A. Seyfarth, Bioinspired balance control assistance can reduce metabolic energy consumption in human walking, TNSRE 27 (9) (2019) 1760–1769.

[132] M.A. Sharbafi, D. Lee, T. Kiemel, A. Seyfarth, Fundamental subfunctions of locomotion, in: Bioinspired Legged Locomotion, Elsevier, 2017, pp. 11–53.

[133] M.A. Sharbafi, A. Seyfarth, G. Zhao, Locomotor sub-functions for control of assistive wearable robots, Front. Neurorobot. 11 (2017) 44.

[134] A. Sarmadi, C. Schumacher, A. Seyfarth, M.A. Sharbafi, Concerted control of stance and balance locomotor subfunctionsleg force as a conductor, IEEE Transact. Med. Robot. Bion. 1 (1) (2019) 49–57.

[135] S.R. Goldberg, S.J. Stanhope, Sensitivity of joint moments to changes in walking speed and body-weight-support are interdependent and vary across joints, J. Biomech. 46 (6) (2013) 1176–1183.

[136] J. Nilsson, A. Thorstensson, Ground reaction forces at different speeds of human walking and running, Acta Physiol. Scand. 136 (2) (1989) 217–227.

[137] M.-C. Chiu, H.-C. Wu, L.-Y. Chang, Gait speed and gender effects on center of pressure progression during normal walking, Gait Post. 37 (1) (2013) 43–48.

[138] S. W. Lipfert, Kinematic and Dynamic Similarities Between Walking and Running, Kovač Hamburg, Germany, 2010.

CHAPTER 8

Stair negotiation made easier using low-energy interactive stairs

Yun Seong Song[a], Sehoon Ha[b], Hsiang Hsu[c], Lena H. Ting[d,e], C. Karen Liu[f]

[a]Department of Mechanical and Aerospace Engineering, Missouri University of Science and Technology, Rolla, MO, United States
[b]Google Brain, Mountain View, CA, United States
[c]Department of Mechanical Engineering, Georgia Institute of Technology, Atlanta, GA, United States
[d]Department of Biomedical Engineering, Georgia Institute of Technology and Emory University, Atlanta, GA, United States
[e]Department of Rehabilitation Medicine, Division of Physical Therapy, Emory University, Atlanta, GA, United States
[f]School of Interactive Computing, Georgia Institute of Technology, Atlanta, GA, United States

1 Introduction

Stair negotiation is a demanding task that limits the independence of individuals with mobility impairments such as muscle weakness, joint pain, or reduced sensorimotor control. Joint moments in the knee are over 3 times greater during stair negotiation compared to level walking during both stair ascent and descent [1–3]. Stair negotiation is ranked among the top 5 most difficult tasks in community-residing older adults [4,5]. Patients—such as those with hip osteoarthritis—adopt altered joint movements to reduce pain during stair negotiation [6]. Moreover, even if they are capable of using stairs, people with mobility impairments often avoid stair negotiation [5,7].

Current solutions providing assistance in stair negotiation are costly and energy consuming, and do not help to retain the user's ability to negotiate stairs independently. Elevators or stairlifts are often impractical to install because they require substantial household remodeling. Further, an elevator can consume over 12,000 kWh annually [8], equivalent to 50% of the average household energy consumption in the United States in 2009 [9] and over 200% of that in the United Kingdom in 2004 [10]. Perhaps more importantly, elevators or stair-lifts replace the need to negotiate stairs altogether, regardless of a user's level of motor function. Because studies suggest that disuse of a specific motor function can further accelerate its loss [11–13],

Powered Prostheses
http://dx.doi.org/10.1016/B978-0-12-817450-0.00008-0

179

it is important to provide motor assistance that allows users to retain their ability to use stairs and to prevent further motor decline.

Cheap, low power, yet effective human movement assistance is possible by applying the principle of energy recycling [14]. Collins et al. showed that a simple exoskeleton with passive springs can store and return energy at each step to assist joint motion during walking [15]. The exoskeleton takes advantage of the alternating braking and propelling action of the legs during gait at the level of the ankle (Fig. 8.1). During braking, the lower–leg exoskeleton stores mechanical energy in a passive spring, reducing the negative work generated by the ankle to brake the leg. The stored energy is later released to propel the body forward, reducing the positive work generated by the ankle for propulsion. Through appropriate detection and response to gait events, the exoskeleton reduces the metabolic cost of walking by 7.2% without consuming energy.

In contrast to level walking, storing, and returning energy within each gait cycle cannot be applied to stair negotiation. The legs produce predominantly positive work during stair ascent, and predominantly negative work during stair descent [2]. Thus a more effective approach for energy-recycling in stair negotiation would be to store a large amount of energy cumulatively during stair descent, and to then release that energy during stair ascent. Specifically, energy recycling could be targeted to two phases

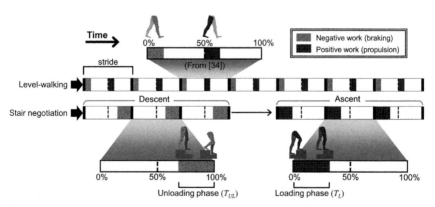

Figure 8.1 *Positive and negative work generation during walking and during stair negotiation.* Dashes represent the phases during which negative work is generated by the right leg (braking). Dots represent the phases during which positive work is generated by the right leg (propulsion). In walking, the right leg both brakes and propels within one stride. In stair negotiation, however, the right leg generates predominantly negative work throughout stair descent (T_{UL}), and predominantly positive work throughout stair ascent (T_L). The gait cycle during stair negotiation follows the definition in [1]. (*Source:* Reproduced with permission from https://doi.org/10.1371/journal.pone.0179637.)

where the greatest increase in joint power are observed compared to level walking: the loading phase of stair ascent (T_L) when energy is *generated* by the leading leg, and the unloading phase of stair descent (T_{UL}) when energy is *absorbed* by the trailing leg (Fig. 8.1). In the loading phase (T_L), center-of-mass (CoM) elevation occurs early in the gait cycle, between foot strike of the leading leg and into the mid swing phase of the trailing leg. Sagittal knee joint power throughout T_L is higher than in level walking, both in adults over 40 years old [3] and healthy young adults [2], with peak knee power of roughly 2 times greater than in level walking [1,2]. In the unloading phase (T_{UL}), the CoM is lowered as the trailing leg does negative work (absorbs energy) until toe-off. In healthy young adults, knee joint power throughout T_{UL} is higher than in level walking [2], and peak knee joint power reaches 3.8 times that seen during level walking [1,2]. Further, the ankle joint generates large negative power in sagittal plane during T_{UL} but negligible negative power during overground walking. Based on previous findings [1–3], we hypothesized that assistive stairs could store energy during T_{UL} and release it during T_L, reducing both positive (ascent) and negative (descent) work generation in the legs during stair negotiation.

Therefore, our objective was to design, build, and test energy-recycling assistive stairs (ERAS) that store energy during stair descent and return it to assist the user during stair ascent. We built two prototype modular steps that can be placed on an existing staircase. Energy is stored in passive springs and released based on gait events detected by pressure sensors on each tread. We then measured the amount of work generated or dissipated by the leg joints during assisted and unassisted stair negotiation in naive users. Our results show that ERAS reduces positive work generation in the leading leg during ascent by 17.4 ± 6.9%. In particular, a 37.7 ± 10.5% reduction was observed in the knee, which is one of the main contributors of positive work during ascent. In addition, ERAS reduces negative work generation in the trailing leg during descent by 21.9 ± 17.8%. In particular, a 26.0 ± 15.9% reduction was observed in the ankle-sagittal degrees of freedom (DOF), which is one of the main contributors of negative work during descent. Together, our work demonstrates the feasibility of a low power, modular, interactive device to assist those with difficulty in stair negotiation in their homes.

2 Materials and methods

2.1 Energy-recycling assistive stairs

Each ERAS module is equipped with its own latch, sensor, and a set of springs (Fig. 8.2). Each is a single stair step designed to be placed on an

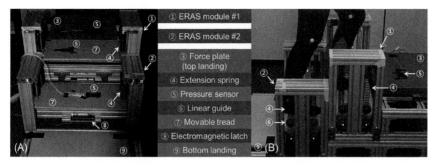

Figure 8.2 *Two ERAS modules with the top and bottom landings:* (A) front view, (B) side view with a user ascending the ERAS. Each ERAS module consists of its own set of extension springs, a pressure sensor, linear guide, movable tread, and an electromagnetic latch. The two modules are operated by a single Arduino board (not shown). A force plate at the top landing measures the ground reaction forces of a human user. (*Source:* Reproduced with permission from https://doi.org/10.1371/journal.pone.0179637.)

existing stair with step height (17 cm) and depth (28 cm). Customizable aluminum frames (80/20® Inc.), encase a movable tread, linear guides, four extension springs, an electromagnetic latch, and a pressure sensor. The tread motion is constrained to be vertical by linear guides with roller bearings (80/20® Inc.). Four extension springs (Century Springs Corp.) connect the movable tread to the frame and are extended as the tread is lowered. ERAS provides three choices of stiffness for each individual spring: 350 N/m, 560 N/m, and 910 N/m. The highest total stiffness is 3640 N/m, which allows users up to 122 kg to use the current ERAS prototype. When the tread is fully lowered, it contacts an electromagnetic latch at the bottom (Docooler H10054, 180 kg holding force). Pressure sensors (Interlink Electronics®) detect foot placement during both ascent and descent, see Video (https://youtu.be/hgwKoo9J410) for ERAS operation.

Based on user feedback in pilot studies, we set the effective spring constant of each ERAS tread to be k_{norm} = 30.80 ± 1.3 N/m per kg of body weight. Using this weight-dependent spring stiffness, η = 26.7 ± 1.1% of the potential energy lost while descending a step height of h = 17 cm $[\Delta E_{potential}$ (J/kg)], is stored in the extended spring $[\Delta E_{spring}$ (J/kg)], such that ΔE_{spring} = 1/2 k_{norm} h^2 = $\eta g h$ = $\eta \Delta E_{potential}$. Note that when the springs are removed or their motion is locked, ERAS modules do not recycle energy (ΔE_{spring} = 0) and are therefore equivalent to a normal set of stairs.

We prepared a staircase for human experiments with a top landing with a force plate, two ERAS, and bottom landing (Figs. 8.2 and 8.3). The first (top) ERAS module was elevated 17 cm above the ground, while the second (bottom) ERAS module was directly on the ground. The 120–cm-long

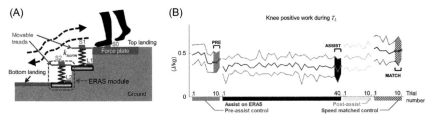

Figure 8.3 *Overview of ERAS human user experiment.* (A) Schematic of the ERAS setup. The pictured compression springs are physically implemented in hardware using extension springs (Fig. 8.2). Participants start each trial on the top landing, storing energy in the springs as they descend the steps. Energy stored in the springs is released back to the user as they ascend the steps. L1 and L2 are the electromagnetic latches, whereas S0, S1, and S2 are the pressure sensors. (B) Positive work generated by the knee over T_L in each trial over an entire experimental session. Each experiment consisted of 10 preassist control trials (*gray*), 40 assist trials (*black*), 10 postassist trials (*dotted*) and 10 speed-matched control trials (*dashed*). Solid lines denote mean and thin lines denote one standard deviation across all participants. (*Source:* Reproduced with permission from https://doi.org/10.1371/journal.pone.0179637.)

bottom landing was 9.2 cm above the ground to match the lowest position of the movable tread on the second ERAS. The 170-cm-long top landing was 43.2 cm above the ground to match the highest position of the movable tread of the first ERAS. A force plate (AccuGait®, Advanced Mechanical Technology, Inc.) formed part of the top landing directly adjacent to the first ERAS. This allowed the measurement of ground reaction forces (GRFs) on the nonmobile top landing instead of on the movable tread, on which the force readings would be affected by and interfere with tread movement. Note that our measures of joint work were restricted to the loading or unloading phases of the gait cycle in the initial (descent) or final (ascent) steps.

The ERAS was designed to be interactive with the user. The electromagnetic latches of the ERAS are modulated based on the inputs from the pressure sensor using a simple binary controller on a single Arduino1 Uno board. Prior to use (home-state), both latches on the first and the second ERAS (L1 and L2, respectively, Fig. 8.3.) are off, with no load detected by the pressure sensors on the force plate, the first ERAS, nor the second ERAS (S0, S1, and S2, respectively, Fig. 8.3). During stair descent, a user steps on S1 which triggers L1 to turn on, locking the movable tread in the lowered position. On the next descending step, the user steps on S2, which then triggers L2 to turn on, latching the next movable tread in the lowered position. During a subsequent stair ascent, stepping on S2 does not trigger an event. On the next ascending step, S1 is pressed which turns L2 off, releasing the movable tread on the second (lower) ERAS. On the next ascending step, S0 is pressed which turns L1 off, releasing the movable tread on the first ERAS. Hence after a stair descent followed by an ascent, the two ERAS are returned back to their home state and ready for the next descent to occur.

The ERAS was also designed to be modular, low cost, and energy efficient. ERAS modules can be customized in size and shape and installed on top of existing staircases. The nonstructural components of a single ERAS unit, that is, sensor, latch, and springs, cost less than $50 and consumes less than 5 W of electricity when a latch is on, and no external power when the latch is off.

2.2 Human experiment

We recruited healthy young participants with no prior knowledge of the ERAS ($n = 9$, 81.0 ± 4.5 kg, 31.1 ± 4.5 years old, 1 female, Table 8.1). All participants provided their written consent to the experiment protocol, which was approved by the Institutional Review Board of Georgia Institute of Technology. All methods, including the experiments, were performed in accordance with the relevant guidelines and regulations of the board.

To measure how participants negotiated stairs without assistance, they first used the ERAS with the treads immobilized, that is, energy-recycling

Table 8.1 Participants' gender, age, weight, total spring constant, and the weight-normalized spring constant of the ERAS used in the experiments.

Participant	Gender	Age (years)	Weight (kg)	Total spring constant (N/m)	Normalized spring constant K_{norm} (N/m/kg)
1	M	30	79	2521.83	31.92
2	M	27	100	2942.13	29.42
3	F	37	46	1401.01	30.46
4	M	25	94	2942.13	31.30
5	M	31	83	2521.83	30.38
6	M	35	85	2521.83	29.67
7	M	33	86	2521.83	29.32
8	M	36	80	2521.83	31.52
9	M	26	76	2521.83	33.18
Mean $\pm$ STD	N/A	31.1 ± 4.5	81.0 ± 15.1	N/A	30.80 ± 1.3

Source: Reproduced with permission from https://doi.org/10.1371/journal.pone.0179637.t001

turned off, for 10 preassist control trials (Fig. 8.3). Next, the springs were connected to the treads which were allowed to move, thereby allowing the springs to store and release energy. Participants were allowed two trials (not analyzed) to familiarize themselves with the operation of the ERAS. Following this, each participant experienced 40 assist trials (30 trials for participant 6) with the springs adjusted to their body weight (k_{norm}, Table 8.1). The springs were then removed in 10 postassist trials to wash out after-effects (if any) from using the ERAS before the next trials began. To ensure comparison across similar stair negotiation speeds when using ERAS [16–18], participants performed 10 additional trials with the springs removed in which they matched their step duration of the last assist trial (speed-matched control trials). The beats per minute of the step cadence in the last assist trial were provided to the participant by and audio beat for approximately 1 minute after the postassist trials. The audio was turned off prior to the speed-matched control trials to prevent participants from marching to the beat. In all other trials, participants self-selected their gait speeds. To avoid averaging transient behaviors over multiple trials during analysis, we selected only the last three preassist control trials (PRE), the last three assist trials (ASSIST), as well as the last three speed-matched control trials (MATCH) for analysis. The blocked conditions were designed to account for the possibility that subject would adapt their stair negotiation strategy on ERAS over repeated exposure, a common phenomenon observed when humans are exposed to novel environmental effects [19–21].

To measure GRFs on the top landing and body segment kinematics, we used a six-axis force plate synchronized with full-body kinematics using a motion-capture system (Vicon®). The force plate provided the GRF and the center of pressure (CoP) at 120 Hz, whereas the motion-capture system provided the subject's full-body kinematics in a motion-capture suit (53 markers), the location of the force plate, and the location of the ERAS tread, also at 120 Hz.

2.2.1 Gait phase definition

We measured joint work for ascent during T_L and that for descent during T_{UL}. The definitions of T_L and T_{UL} follow the gait cycle segmentation in [1] (Fig. 8.4). During ascent, T_L was identified as the time interval that began with the leading foot contact on the force plate (0% of the gait cycle) and ended as the trailing foot was lifted to the height of the leading foot (★32%), equivalent to the combination of "Weight Acceptance" and "Pull Up" phases [1]. During descent, T_{UL} was identified as the time interval that began at the second mid-swing of a gait cycle (*70%) and ended with the trailing foot toe-off (100%), equivalent to the "Controlled Lowering" phase defined in [1]. Measurements from the force plate and the motion–capture system were used to identify T_L and T_{UL}.

To test whether using ERAS affected stair negotiation outside of T_L or T_{UL}, we defined the time interval T_{FCA} (where FC stands for forward-continuance [1] and A stands for ascent) and T_{FCD} (where D stands for descent, Fig. 8.4). T_{FCA} began at the end of T_L (at ~32% [1]) and ended at the end of the subsequence double-stance (63.6% [2]), making $| \ T_{FCA} \ | \simeq | \ T_L \ |$. Note that T_{FCA} is equivalent to the "Forward Continuance" phase during ascent defined in Ref. [1]. T_{FCD} began at the foot contact (38.8% [2]) and ended at the beginning of T_{UL} (~70% [1]) making $| \ T_{FCD} \ | \simeq | \ T_{UL} \ |$. Note that T_{FCD} is equivalent to the combination of "Weight Acceptance" and "Forward Continuance" phases during decent defined in Ref. [1]. During T_{FCA} and T_{FCD}, the treads on the ERAS were not moving since there were no feet placed on the treads. With our measurement, we identified T_{FCA} and T_{FCD} by identifying the beginning of T_{FCA} and the end of T_{FCD} from foot markers and using the fact that $| \ T_{FCA} \ | \simeq | \ T_L \ |$ and $| \ T_{FCD} \ | \simeq | \ T_{UL} \ |$.

2.2.2 Joint work calculation

To calculate joint moments and velocities and eventually the joint work, we used inverse dynamics using an open source physics engine, DART [22,23]. We computed the joint moment and velocity from the recorded kinematics, GRF, and CoP. Joint positions q were obtained by solving a standard inverse kinematics problem. Next, we derived the joint velocities $\dot{q}$ and

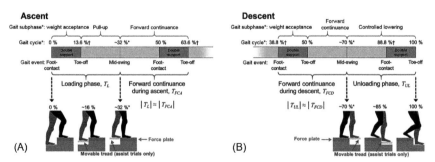

Figure 8.4 *Gait phases during stair negotiation.* (A) During ascent, T_L begins with the leading foot contact (0%) and ends at mid-swing of the trailing leg (~32%*). The forward-continuance phase, T_{FCA}, begins at mid-swing and ends at the end of the double-support phase (63.6%†). (B) During descent, T_{UL} begins at the second mid-swing of the leading leg (~70%*) and ends at the trailing leg toe-off (100%). The forward-continuance phase, T_{FCD}, begins with the foot contact (38.8%†) and ends at the mid-swing. *As defined in Ref. [1]. †As defined in Ref. [2]. (*Source:* Reproduced with permission from https://doi.org/10.1371/journal.pone.0179637.)

accelerations $\ddot{q}$ by taking fourth-order central finite differences of the joint trajectories. Finally, we obtained the joint moments τ from the equations of motion: $M(q)\ddot{q} + C(q,\dot{q}) = \tau + J^T F$, where q is joint positions, M is the inertia matrix, C is the Coriolis and gravity vector calculated by DART, J is a Jacobian matrix, and F is the GRF. The mass distribution of participants was assumed to follow the adapted inertial parameters in the study by DeLeva [24].

Joint power P was calculated by taking the inner product of the instantaneous joint moment and velocity vectors of the leading leg joints during ascent and the trailing leg joints during descent [16]. Positive work in a single DOF during T_L was obtained by taking the integral of P when it is positive during T_L, denoted by the domain POS, such that (1) $W^+_{joint} = \int_{POS} P\,dt$, where W^+_{joint} is the positive work of a particular joint. Similarly, negative work in a single joint during T_{UL} was obtained by taking the integral of P only when it is negative during T_{UL}, denoted by the domain NEG, such that (2) $W^-_{joint} = \int_{NEG} P\,dt$, where W^-_{joint} is the negative work of a particular joint. To quantify the assistance provided by the ERAS, we obtained individual joint positive work as well as the total positive work generation from all joint DOFs during T_L, and individual joint negative work as well as the total negative work generation from all joint DOFs during T_{UL} (Table 8.2). Specifically,

Table 8.2 Work metrics and their definitions.

Metric	Description	Obtained during
W^+_{TOT}	Total positive work generated by the leading leg in all DOF	Ascent, T_L
W^+_{Sag}	Total positive work generated by the leading leg in sagittal plane	Ascent, T_L
W^+_{hip}	Positive work generated by the leading-leg hip joint in sagittal plane	Ascent, T_L
W^+_{knee}	Positive work generated by the leading-leg knee joint	Ascent, T_L
W^+_{ank}	Positive work generated by the leading-leg ankle joint in sagittal plane	Ascent, T_L
W^-_{TOT}	Total negative work generated by the trailing leg in all DOF	Descent, T_{UL}
W^-_{Sag}	Total negative work generated by the trailing leg in sagittal plane	Descent, T_{UL}
W^-_{hip}	Negative work generated by the trailing-leg hip joint in sagittal plane	Descent, T_{UL}
W^-_{knee}	Negative work generated by the trailing-leg knee joint	Descent, T_{UL}
W^+_{ank}	Negative work generated by the trailing-leg ankle joint in sagittal plane	Descent, T_{UL}

Source: Reproduced with permission from https://doi.org/10.1371/journal.pone.0179637.t002

we calculated the positive work generated by each of the five DOFs of the leading leg (hip-sagittal, hip-frontal, knee, ankle-sagittal, and ankle-frontal joints) during T_L. Then, we summed up the positive work generated by these five DOFs to find the total positive work generated by the leading leg during T_L (W_{TOT}^+). In addition, we summed the positive work generated by the three sagittal DOFs (i.e., by the hip-sagittal, knee, and the ankle-sagittal DOFs) to find positive work generation in the sagittal plane only (W_{Sag}^+). We also calculated the negative work generated by each of the five DOFs of the trailing leg during T_{UL}, the total negative work (W_{TOT}^-), as well as the negative work generation in the sagittal plane only (W_{Sag}^-). We repeated these calculations in three conditions: PRE, ASSIST, and MATCH.

2.2.3 Statistical analysis

To quantify how much joint work was reduced by using ERAS, we compared the total positive or negative work, the work by all sagittal DOFs, as well as the work by individual DOFs. We compared the work metrics in three conditions (PRE, ASSIST, and MATCH) and nine participants using ANOVA, with both "conditions" and "participants" as fixed factors. Tukey HSD was used to test the mean differences among conditions with significance at $P < 0.05$.

3 Results

3.1 Operation of energy-recycling stairs

The ERAS stores energy during the unloading phase (T_{UL}) in stair descent and subsequently returns energy to the user during the loading phase (T_L, Figs. 8.1 and 8.4) in stair ascent. During T_{UL}, CoM lowering begins as the leading leg pushes the movable tread down and stores energy in the extended springs (Fig. 8.4B at ~70%). Energy storage in the springs is complete when the leading foot (and therefore also the leading movable tread) is fully lowered by the step height (Fig. 8.4B between 88.8% and 100%). As a result, T_{UL} fully encompasses the time interval during which the storage of energy in the springs would affect negative work during stair descent. During T_L, CoM elevation begins with the leading foot contact, which triggers the release of energy from the springs of the lower tread to the trailing leg. Energy release begins shortly after as the trailing foot begins to be elevated (Fig. 8.4A between 0% and 13.6%). Energy release is complete when the trailing foot (and therefore also the trailing movable tread) is fully elevated by the step height (Fig. 8.4A at ~32%). As a result, T_L fully encompasses the time interval during which the release of energy in the springs affects positive work during gait ascent (see Video 8.1).

3.2 Assessment of assistance provided during stair negotiation

All users—ranging from 46 to 100 kg (Table 8.1)—were successful in stair negotiation on the ERAS. Participants had no prior knowledge about the ERAS, nor were they provided with any information about the purpose of the ERAS during the experiment session. In over 360 trials across nine participants, no adverse events or concerns about safety were reported.

During ascent, the step duration in the ASSIST condition was significantly longer than the step duration in the PRE condition (ASSIST vs. PRE, $P < 0.001$), but was not different from the step duration in the MATCH condition (ASSIST vs. MATCH, $P > 0.5$, Fig. 8.5A). During descent, the step duration during the ASSIST condition was not different from the step duration in the PRE condition (ASSIST vs. PRE, $P > 0.5$), but was significantly shorter than the step duration in the MATCH condition (MATCH vs. ASSIST, $P < 0.001$, Fig. 8.5B). In other words, in the AS-SIST condition, stair ascent was slower than during normal stair negotiation (PRE) but descent was at a comparable speed. This counter-intuitive

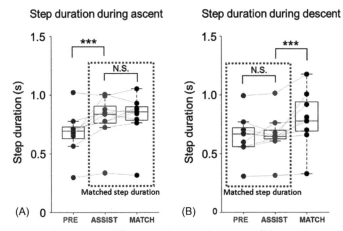

Figure 8.5 *Step duration in different stair negotiation conditions.* (A) During ascent, step duration on the ERAS was significantly longer than in normal stair negotiation (ASSIST vs. PRE). No significant difference was observed in the step duration between ERAS and normal stairs with slower, matched gait speed (ASSIST vs. MATCH). Thus we compared the results (ASSIST) against MATCH (instead of PRE) during ascent. (B) During descent, step duration was not different on ERAS vs. during normal stair negotiation (ASSIST vs. PRE). However, the descent steps in the MATCH condition were significantly longer than on ERAS (ASSIST vs. MATCH). Thus we compared the results (ASSIST) against PRE (instead of MATCH). *** refers to $P < 0.001$, * refers to $P < 0.05$, and N.S. refers to no significant difference. (*Source:* Reproduced with permission from https://doi.org/10.1371/journal.pone.0179637.)

observation is discussed later in "Discussion." Based on step duration, we used the MATCH condition as our speed-matched control for the AS-SIST condition during ascent, and used the PRE condition as our speed-matched control for the ASSIST condition during descent.

ERAS did not qualitatively affect movement kinematics or CoM motion, consistent with self-reports of the ERAS being easy to use. Fig. 8.6 shows the joint angle and CoM trajectories of a representative participant (#9). Sagittal plane joint angles of the hip, knee, and ankle were qualitatively similar during T_L between the ASSIST and the MATCH conditions, with the hip-sagittal angle slightly higher and the ankle-sagittal angle slightly lower than in MATCH. The CoM height from the ground also followed similar trajectories between ASSIST and MATCH conditions (ascent) as well as between ASSIST and PRE conditions (descent). There was no evidence of sudden lifts or drops of CoM.

3.2.1 Stair ascent

In speed-matched normal stair ascent (MATCH), the total positive work of 1.317 ± 0.134 J/kg was generated by the leading leg during T_L (W^+_{TOT}, Fig. 8.7A). The sagittal-plane DOFs (hip-sagittal, knee, and ankle-sagittal) generated over $95.8 \pm 2.1\%$ (W^+_{Sag}, 1.262 ± 0.137 J/kg) of the total positive work. The hip-sagittal DOF generated $51.4 \pm 8.1\%$ (W^+_{hip}, 0.677 ± 0.107 J/kg), the knee DOF generated $40.0 \pm 7.9\%$ (W^+_{knee}, 0.527 ± 0.104 J/kg), and the ankle-sagittal DOF generated $4.2 \pm 2.7\%$ (W^+_{ank}, 0.056 ± 0.036 J/kg) of the total positive work during T_L. Contribution of frontal DOFs (hip-frontal and ankle frontal) to total positive work was $4.4 \pm 2.0\%$.

Using ERAS significantly reduced positive work generation during T_L. When speed was matched, using ERAS reduced W^+_{TOT} by $17.4 \pm 6.9\%$ (AS-SIST vs. MATCH, $P < 0.001$, Fig. 8.7A). W^+_{knee} was reduced by $37.7 \pm 10.5\%$ (ASSIST vs. MATCH, $P < 0.001$). However, W^+_{hip} did not change significantly (ASSIST vs. MATCH, $P > 0.25$).

During the forward-continuance phase of ascent, T_{FCA}, positive joint work generated by each DOF showed no difference between ASSIST and MATCH conditions, except for the hip-sagittal DOF, which generated 0.142 ± 0.063 J/kg more positive work during ASSIST than in MATCH (Fig. 8.8). However, because the total positive work reduction during T_L was 0.230 ± 0.094 J/kg, the total positive work during $T_L + T_{FCA} = (0\%, 64\%)$ showed a trend of reduction by 0.088 ± 0.108 J/kg ($P < 0.2$).

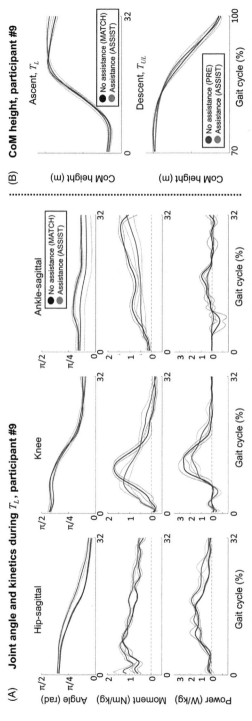

Figure 8.6 *Joint kinematics and center-of-mass trajectories of participant #9.* (A) Joint angle, moment, and power for the hip-sagittal, knee and ankle-sagittal DOFs during T_L (0%–32% of the gait cycle during ascent) for ASSIST (*gray*) and MATCH (*black*) conditions. The solid line indicates the mean trajectory and the dashed lines indicate one standard deviation. (B) Center-of-mass height over time during T_L (top), and during T_{UL} (bottom). *Dark gray* trajectories indicate PRE condition. (*Source:* Reproduced with permission from https://doi.org/10.1371/journal.pone.0179637.)

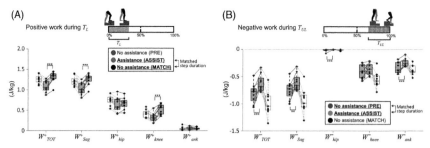

Figure 8.7 *Positive and negative work on ERAS.* Work metrics in the ASSIST condition (shaded boxes, *gray* dots) and in speed-matched control conditions (MATCH for ascent, PRE for descent). Thin and dark gray lines connect each participant's result in *dark gray, gray,* and *black* dots. (A) Ascent: Between ASSIST and MATCH conditions, W^+_{TOT}, W^+_{Sag}, and W^+_{knee} were reduced by 17.4 ± 6.9%, 17.8 ± 7.3%, and 37.7 ± 10.5%, respectively. PRE trials are also shown for reference (*white* boxes, *dark gray* dots). (B) Descent: Between ASSIST and PRE conditions, W^-_{TOT}, W^-_{Sag}, and W^-_{ank} were reduced by 21.9 ± 17.8%, 16.9 ± 21.3%, and 26.0 ± 15.9%, respectively. W^-_{hip} was also significantly reduced, but the absolute reduction is very small. MATCH trials are also shown for reference (*white* boxes, *black* dots). Significance of $P < 0.001$ are noted as *** and $P < 0.01$ are noted as **. (*Source:* Reproduced with permission from https://doi.org/10.1371/journal.pone.0179637.)

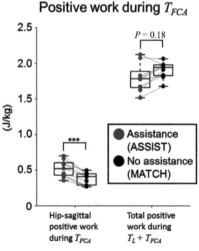

Figure 8.8 *Positive work generated during T_{FCA}.* Positive work generated by the hip-sagittal DOF was significantly higher in the ASSIST condition (0.523 ± 0.119 J/kg) versus the MATCH condition (0.381 ± 0.084 J/kg, $P < 0.001$). The total positive work generated during $T_L + T_{FCA}$ was not significantly different, with a trend toward slight reduction in ASSIST from MATCH (0.088 ± 0.101 J/kg, $P < 0.2$). (*Source:* Reproduced with permission from https://doi.org/10.1371/journal.pone.0179637.)

3.2.2 Stair descent

In speed-matched normal stair ascent (PRE), the total negative work of -0.807 ± 0.179 J/kg was generated by the leading leg during T_{UL} (W^-_{TOT}, Fig. 8.7B). The sagittal-plane DOFs (hip-sagittal, knee and ankle-sagittal) generated $92.2 \pm 4.2\%$ (W^-_{Sag}, -0.744 ± 0.169 J/kg) of the total negative work. The knee DOF generated $48.7 \pm 18.8\%$ (W^-_{knee}, -0.393 ± 0.152 J/kg), the ankle-sagittal DOF generated $41.5 \pm 15.3\%$ (W^-_{ank}, -0.335 ± 0.123 J/kg), and the hip-sagittal DOF generated $1.9 \pm 3.4\%$ (W^-_{hip}, -0.016 ± 0.028 J/kg) of the total negative work during T_{UL}. Contribution of frontal DOFs (hip-frontal and ankle-frontal) to total negative work was $7.9 \pm 5.0\%$.

Using ERAS significantly reduced negative work generation during T_{UL}. When speed was matched, using ERAS reduced W^-_{TOT} by $21.9 \pm 17.8\%$ (PRE vs. ASSIST, $P < 0.001$, Fig. 8.7B). W^-_{ank} was reduced by $26.0 \pm 15.9\%$ (PRE vs. ASSIST, $P < 0.001$). However, W^-_{knee} did not change significantly (PRE vs. ASSIST, $P > 0.68$).

Using ERAS did not change the total joint work during the forward-continuance phase during descent, T_{FCD}. Negative work generated by each DOFs, as well as the total of them, were not different between PRE and ASSIST conditions.

4 Discussion

Our ERAS recycle energy from human movement to assist stair negotiation without the use of high-power actuators. Naive users with no prior exposure to ERAS were able to safely and effectively use our stairs, which operates with a system of springs and movable treads. We show that ERAS stores energy that is usually absorbed by the trailing leg during stair descent. Although further investigation is required, this reduction in negative work could possibly reduce the metabolic cost of stair descent, and also reduce muscle and joint forces that cause pain [25,26]. Similarly, stored energy in the ERAS is returned to users during stair ascent, showing a trend toward reducing the amount of positive work generated by the user. This could also lead to reduce metabolic energy as well as muscle and joint forces during stair ascent.

4.1 Implications of ERAS as an assistive device

ERAS is the first nonwearable device that recycles human-generated mechanical energy to provide motor assistance (Table 8.3). Several existing wearable energy-recycling devices convert mechanical energy stored from

Table 8.3 Current energy-recycling devices for humans.

	Nonwearable	Wearable
Output: mechanical energy for human movement assistance	ERAS (this work)	Collins 2015 [15]
Output: electrical energy for external device operations	Pavegen®[27]	Niu 2004 [28], Hayashida 2000 [29], Paradiso 2005 [30], Riemer 2010 [31], Donelan 2008 [14], Rome 2005 [32], Granstrom 2007 [33],

Energy-recycling devices in the literature are categorized based on their configuration (wearable vs. nonwearable) and energy output (mechanical vs. electrical energy). The underlined device is for stair negotiation; all other devices are for level walking.

Source: Reproduced with permission from https://doi.org/10.1371/journal.pone.0179637.t003

human movement into electrical energy to power external devices [14, 27–35]. However, converting mechanical energy to electrical energy is inefficient and unnecessary for applications requiring mechanical energy output for movement assistance. Wearable devices [14,15,28–33] are constrained in the amount of energy that can be stored and released due to tight constraints on space and weight. By installing the energy recycling mechanism on a staircase, the ERAS can use large (and possibly heavier) springs, which allow for a greater amount of energy storage, without the form factor constraints imposed by wearable devices. As user compliance in wearing assistive devices can be limited, modifications in the home environment may be more beneficial in providing greater independence, safety, and mobility to those with difficulty with stair negotiation. Moreover, while all other existing energy-recycling devices available target overground human locomotion such as walking or running, ERAS is the only device tailored for assistance during stair negotiation.

For people with reduced motor function who can still negotiate stairs with some assistance, our modular and low-power ERAS is a low-cost and effective alternative to the existing high-power and expensive options such as elevators, escalators or stair lifts. The modular design of ERAS allows quick and easy installation (and removal, if necessary) on top of an existing staircase without expensive remodeling. The passive springs, low-power electromagnetic latches, and a single Arduino board together require very little power compared to motorized devices, thereby reducing the cost of use. The low-cost and low-power design allows ERAS to be more affordable and practical to people with limited financial resources. The modular ERAS can easily integrate onto existing stairs, making homes, and communities more suited for aging-in-place [36].

One surprising outcome of ERAS is its ability to assist not only stair ascent, but also stair descent. Perhaps counter-intuitively, stair descent can be as energy demanding as stair ascent. A person (body weight = mg) ascending a step (height = h) must generate net positive energy in the legs of mgh. Similarly, descending a step requires a potential energy of mgh to be dissipated by negative work generation in the legs and through collision. Hence, if energy lost through collision is small, the negative work during descent can be of similar magnitude to positive work in ascent. In fact, Fig. 8.7 shows that W_{TOT}^- is comparable in magnitude to W_{TOT}^+. Moreover, the control of muscle braking is more challenging than generating positive work with muscles, and is a focus of exercises to help frail older adults to better descend stairs and to decrease the risk of fall [37].

While the current hardware prototype allows initial investigation of the assistance provided by ERAS, future evolutions of ERAS could allow higher usability in practical settings. To use the current prototype, a user needs to first descend and then ascend the ERAS modules in order to store and then to return mechanical work. This constraint can be eliminated in the future prototypes by incorporating additional mechanisms to configure the ERAS to either "ascend-able" (treads down) or "descend-able" (treads up) positions, thereby also allowing multiple users to consecutively ascend or descend the ERAS modules.

4.2 Experimental limitations and justifications

The mismatch in the descent step duration between ASSIST and MATCH conditions (Fig. 8.5) may be due to the verbal instructions to the participants as well as the tempo of the audio beat. We informed the participants that they might walk more slowly in the following trials than they would normally on normal stairs. This may have caused some participants to be predisposed to slow down during both ascent and descent compared to PRE trials. However, since stair descent was not slower in ASSIST than in PRE, stair descent in MATCH turned out to be slower than in ASSIST. Also, the audio beat we selected was similar to the speed of ascent of the last ASSIST trial, but was too slow for the descent. Speed-match trials in future experiments should be conducted with refined instructions and separate audio beat tempos for ascent and descent to more accurately represent the step durations during ASSIST.

We only examined joint work as a first step, but further study on the specific effects of ERAS on more physiologically relevant metrics is warranted, particularly in mobility-limited individuals. These could include joint range of motion [6], peak joint moment/power [1–3], intraarticular joint forces [38], muscle coactivation [39,40], net metabolic energy reduction [15], or the amount of eccentric muscle contraction required [41]. As a first step, we assessed the efficacy of the ERAS in providing assistance using measures of joint work. Although our metrics demonstrate that mechanical work was stored and returned to the user, it is still possible that participants coactivated muscles to stiffen the joints when using the ERAS, which could increase metabolic energy, muscle forces, and interjoint forces; it is not clear whether such changes would increase or reduce joint pain. Factors specific to mobility limitations in specific populations also need to be directly studied when considering the appropriateness of ERAS in providing assistance in stair negotiation.

We were unable to measure the reaction forces on the movable stair treads, but believe that measures from the top landing are comparable to those from intermediate steps on the ERAS. Gait patterns are similar in stair ascent when stepping on the top landing as when stepping onto the intermediate ERAS steps with the tread in the lowered position. Similarly, in stair descent lowering the body from the top landing is similar to lowering the body on the ERAS with the tread locked in the lowered position. Although the ideal case would be to measure GRFs on the moving treads, this would require very light and thin force plates, or for the ERAS modules to be isolated and mounted on stationary force plates [2].

5 Summary

Our promising results that show energy-recycling during stair negotiation in young healthy participants motivate further refinement and optimization of the ERAS to aid older adults and individuals with a wide range of mobility impairments. As healthy users could safely benefit from ERAS without explicit instructions or training, more effective user guidelines and practice could facilitate stair negotiation for those with muscle weakness or joint pain. The physical design of the ERAS could be tailored to provide more user-specific trajectories of energy storage and release. Further reduction in the net electrical energy expenditure could also be achieved through a more refined mechanical design and well as the harvesting energy from stair motion to power the system's electronics. It is possible that each ERAS could be operated independently without external power. In addition, future ERAS could overcome the limitations of the current prototype, such as to provide assistance to multiple users. Overall, our proof-of-principle demonstration provides a novel platform for interactive, personalized, energy-efficient, and cost-effective devices for assisting stair negotiation to suit a wide range of individuals with reduced mobility.

Acknowledgments

This work was supported by the National Science Foundation Grant EFRI-1137229. The authors would like to thank Dr. Young-Hui Chang and Dr. Randy Trumbower for use of their equipment.

References

[1] B.J. McFadyen, D.A. Winter, An integrated biomechanical analysis of normal stair ascent and descent, J. Biomech. 21 (9) (1988) 733–744.

[2] R. Riener, M. Rabuffetti, C. Frigo, Stair ascent and descent at different inclinations, Gait Posture 15 (1) (2002) 32–44.

[3] S. Nadeau, B.J. McFadyen, F. Malouin, Frontal and sagittal plane analyses of the stair climbing task in healthy adults aged over 40 years: what are the challenges compared to level walking?, Clin. Biomech. 18 (10) (2003) 950–959.

[4] J.D. Williamson, L.P. Fried, Characterization of older adults who attribute functional decrements to old age, J. Am. Geriat. Soc. 44 (12) (1996) 1429–1434.

[5] J. Startzell, D. Owens, L. Mulfinger, P. Cavanagh, Stair negotiation in older people: a review, J. Am. Geriatr. Soc. 48 (5) (2000) 567–580.

[6] M. Hall, T.V. Wrigley, C.O. Kean, B.R. Metcalf, K.L. Bennell, Hip biomechanics during stair ascent and descent in people with and without hip osteoarthritis, 2016.

[7] M. Morlock, E. Schneider, A. Bluhm, M. Vollmer, G. Bergmann, V. Müller, et al. Duration and frequency of every day activities in total hip patients, J. Biomech. 34 (7) (2001) 873–881.

[8] ThyssenKrupp elevator energy consumption calculator. Available from: https://www.thyssenkruppelevator.com/Tools/energy-calculator.

[9] US Energy Information Administration Annual Energy Review. Available from: http://www.eia.gov/totalenergy/data/annual/pdf/sec2_17.pdf, 2011.

[10] S. Firth, K. Lomas, A. Wright, R. Wall, Identifying trends in the use of domestic appliances from household electricity consumption measurements, Ener. Build. 40 (5) (2008) 926–936.

[11] S.A. Bloomfield, Changes in musculoskeletal structure and function with prolonged bed rest, Med. Sci. Sports Exer. 29 (2) (1997) 197–206.

[12] S.C. Bodine, Disuse-induced muscle wasting, Int. J Bioch. Cell Biol. 45 (10) (2013) 2200–2208.

[13] D.R. Dolbow, A.S. Gorgey, Effects of use and disuse on non-paralyzed and paralyzed skeletal muscles, Aging Dis. 7 (1) (2016) 68.

[14] J.M. Donelan, Q. Li, V. Naing, J. Hoffer, D. Weber, A.D. Kuo, Biomechanical energy harvesting: generating electricity during walking with minimal user effort, Science 319 (5864) (2008) 807–810.

[15] S.H. Collins, M.B. Wiggin, G.S. Sawicki, Reducing the energy cost of human walking using an unpowered exoskeleton, Nature 522 (7555) (2015) 212–215.

[16] J.M. Donelan, R. Kram, A.D. Kuo, Simultaneous positive and negative external mechanical work in human walking, J. Biomech. 35 (1) (2002) 117–124.

[17] T. Keller, A. Weisberger, J. Ray, S. Hasan, R. Shiavi, D. Spengler, Relationship between vertical ground reaction force and speed during walking, slow jogging, and running, Clin. Biomech. 11 (5) (1996) 253–259.

[18] R.L Routson, The Effects of varying speed on the biomechanics of stair ascending and descending in healthy young adults: inverse kinematics, inverse dynamics, electromyography and a pilot study for computational muscle control and forward dynamics, The Ohio State University, 2010.

[19] R. Shadmehr, F.A. Mussa-Ivaldi, Adaptive representation of dynamics during learning of a motor task, J. Neurosci. 14 (5) (1994) 3208–3224.

[20] D.S. Reisman, R. Wityk, K. Silver, A.J. Bastian, Locomotor adaptation on a split-belt treadmill can improve walking symmetry post-stroke, Brain 130 (7) (2007) 1861–1872.

[21] T.D. Welch, L.H. Ting, Mechanisms of motor adaptation in reactive balance control, PLoS One 9 (5) (2014) e96440.

[22] DART. Dynamic Animation and Robotics Toolkit, Available from: http://dartsim.github.io/, 2016.

[23] C.K. Liu, S. Jain, A short tutorial on multibody dynamics, GIT-GVU-15-01-1, Georgia Institute of Technology, School of Interactive Computing, 2012.

[24] P. De Leva, Adjustments to Zatsiorsky-Seluyanov's segment inertia parameters, J. Biomech. 29 (9) (1996) 1223–1230.

[25] P. DeVita, T. Hortobagyi, Age causes a redistribution of joint torques and powers during gait, J. Appl. Physiol. 88 (5) (2000) 1804–1811.

[26] K.R. Kaufman, C. Hughes, B.F. Morrey, M. Morrey, K.N. An, Gait characteristics of patients with knee osteoarthritis, J. Biomech. 34 (7) (2001) 907–915.

[27] Pavegen. Available from: http://www.pavegen.com/.

[28] P. Niu, P. Chapman, R. Riemer, X. Zhang, Evaluation of motions and actuation methods for biomechanical energy harvesting. in: Power Electronics Specialists Conference, 2004. PESC 04. 35th Annual. Vol. 3., IEEE; 2004. pp. 2100–2106.

[29] J.Y., Hayashida, Unobtrusive integration of magnetic generator systems into common footwear. MIT Media Lab, 2000.

[30] J.A. Paradiso, T. Starner, Energy scavenging for mobile and wireless electronics, IEEE Pervas. Comput. 4 (1) (2005) 18–27.

[31] R. Riemer, A. Shapiro, S. Azar, Optimal gear and generator selection for a knee biomechanical energy harvester. in: 1st International Conference on Applied Bionics and Biomechanics, October 2010, p. 14–16.

[32] L.C. Rome, L. Flynn, E.M. Goldman, T.D. Yoo, Generating electricity while walking with loads, Science 309 (5741) (2005) 1725–1728.

[33] J. Granstrom, J. Feenstra, H.A. Sodano, K. Farinholt, Energy harvesting from a backpack instrumented with piezoelectric shoulder straps, Smart Mater. Struct. 16 (5) (2007) 1810.

[34] R. Riemer, A. Shapiro, Biomechanical energy harvesting from human motion: theory, state of the art, design guidelines, and future directions, J. Neuroeng. Rehabilit. 8 (1) (2011) 1.

[35] E. Schertzer, R. Riemer, Harvesting biomechanical energy or carrying batteries? An evaluation method based on a comparison of metabolic power, J. Neuroeng. Rehabilit. 12 (1) (2015) 1.

[36] J. Pynoos, R. Caraviello, C. Cicero, Lifelong housing: the anchor in aging-friendly communities, Generations 33 (2) (2009) 26–32.

[37] P.C. LaStayo, G.A. Ewy, D.D. Pierotti, R.K. Johns, S. Lindstedt, The positive effects of negative work: increased muscle strength and decreased fall risk in a frail elderly population, J. Gerontol. Series A Biol. Sci. Med. Sci. 58 (5) (2003) M419–M424.

[38] P. McLaughlin, P. Chowdary, R. Woledge, A. McCarthy, R. Mayagoitia, The effect of neutral-cushioned running shoes on the intra-articular force in the haemophilic ankle, Clin. Biomech. 28 (6) (2013) 672–678.

[39] J.D. Childs, P.J. Sparto, G.K. Fitzgerald, M. Bizzini, J.J. Irrgang, Alterations in lower extremity movement and muscle activation patterns in individuals with knee osteoarthritis, Clin. Biomech. 19 (1) (2004) 44–49.

[40] S.J. Preece, R.K. Jones, C.A. Brown, T.W. Cacciatore, A.K. Jones, Reductions in co-contraction following neuromuscular re-education in people with knee osteoarthritis, BMC Musculoskeletal Dis. 17 (1) (2016) 372.

[41] P.C. LaStayo, J.M. Woolf, M.D. Lewek, L. Snyder-Mackler, T. Reich, S.L. Lindstedt, Eccentric muscle contractions: their contribution to injury, prevention, rehabilitation, and sport, J. Orthop. Sports Phys. Ther. 33 (10) (2003) 557–571.

CHAPTER 9

Semi-active prostheses for low-power gait adaptation

Peter Gabriel Adamczyk

University of Wisconsin-Madison, Madison, WI, United States

1 Adaptability in lower-limb control

Mobility impairments due to lower-limb amputation affect millions of people worldwide. Prostheses are the enabling technologies in the life-enhancing effort to restore locomotion (standing, walking and running), for the physical and emotional health and personal and economic self-sufficiency of these individuals.

Traditional lower-limb prostheses are passive devices—structures built of different combinations of materials to approximate the biomechanical behavior of the leg and its joints. Such passive systems offer advantages such as simplicity, low weight and height, and modest cost. Their major disadvantage is that they have static mechanical properties, which reduces adaptability across multiple tasks [1]—a substantial shortcoming compared to the natural ankle's ability to adapt broadly to different conditions through variation of stiffness, posture, and ankle energy output [2-4]. Active robotic prostheses offer the potential to replace a wider range of natural function, but also have drawbacks in weight, height, cost, power demand, and control complexity [5-8].

Semi-active prostheses offer an opportunity to combine the benefits of passive and active devices to provide a solution that adapts to different task demands, yet still has low weight, height, complexity, and cost (Fig. 9.1). The semi-active approach allows the user to provide body-scale power input and control using intact and residual structures, while the prosthesis adapts its properties to provide a beneficial mechanical interface to the environment.

1.1 Current semi-active prostheses

The lower-limb prosthetics industry already embraces some aspects of semi-active design. Most current microprocessor-controlled components fall into this category. For example, the Ottobock *C-Leg*, *Genium* and *X3*

Powered Prostheses
http://dx.doi.org/10.1016/B978-0-12-817450-0.00009-2

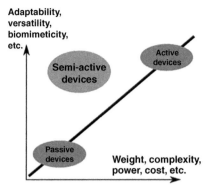

Figure 9.1 Goal of semi-active devices is to achieve many of the benefits of active devices, with a low user burden closer to passive devices.

knee modules and the Össur Rheo knee all use controlled hydraulic damping to adapt to different tasks and terrains [9-12]. Several foot-ankle systems also use variable hydraulic dampers, including the Endolite *Elan* [13], College Park *Odyssey* [14,15] Fillauer *Raize* [16], Ottobock *Meridium* [17] and *Triton Smart Ankle* [18], and Freedom Innovations *Kinnex* [19]. The Össur *Proprio* foot [20] actively adapts the ankle angle during swing phases under robotic control.

Other research prostheses have controlled foot shape [21], stiffness [22-24], or angle [6] using different mechanisms. But, no consistent framework has been described to relate the different approaches and compare their strengths and weaknesses.

1.2 Benefits of modulating properties in prostheses

These current semi-active prostheses, and other approaches used in research, have substantially different biomechanical consequences. For example, a hydraulic ankle that dorsiflexes by motion in a damper during stance phase to adapt to sloped ground [25] dissipates much of the energy that would otherwise be returned by a compliant foot member [1,26,27]. In contrast, a rigid ankle that dorsiflexes robotically during swing [20] dissipates no energy from the body while accomplishing the same goal. Furthermore, a prosthesis that lowers its stiffness to allow the same dorsiflexion may actually store and return *more* energy than the robotic version [28]. On the other hand, the hydraulic ankles are excellent for promoting smooth motion, which may be just as valuable to the user as energy return [1]. Similar or complementary discrepancies will occur in other movements of gait as well, with potentially important effects on the economy and biomechanical loading required of the user.

These examples are only a small sample of the possibilities for enhancing adaptability using semi-active prosthetic design. A wide range of biomimetic concepts, design approaches and technological principles could potentially be applied to adapt prostheses to different aspects of human locomotion and behavior. This chapter outlines a theoretical, experimental and design-oriented framework for mimicking different behaviors of the ankle-foot system using semi-actively controlled devices. Examples are presented that pursue shape, stiffness, and ankle angle: three controllable parameters that can influence biomechanics during walking and standing.

2 The semi-active approach to biomechatronic design

Opportunities to create semi-active systems are not straightforward to identify. Most of what we observe about human motion involves both force and motion of the limbs and tissues—the full dynamic action of the biomechanical system. Systems that create motion in (or opposite to) the direction of an applied force are inherently performing work; hence, much of what we observe about movement seems dependent on the production, dissipation, storage and return of mechanical power. At first glance, then, the idea of mimicking human motion without providing power seems impossible. However, closer inspection reveals substantial subsets of these behaviors that can be mimicked by passive elements.

First, much of the work-involved movement at many joints involves negative work—the joints resist movements that are applied externally or result from the body's inertia. For example, heel-strike in walking is followed by a phase of resisted plantarflexion and knee flexion, in which the ankle dorsiflexor and knee extensor muscles absorb energy to slow the body's downward fall. These energy-dissipating portions of natural movement are relatively easy to mimic with a range of techniques from passive foam to controlled brakes and hydraulic dampers.

Second, another significant portion of the work-involved movement involves cyclical absorption and generation of work at individual joints or across two joints such as the ankle and knee. For example, the simple act of stabilizing the ankle in standing is mediated by muscle activity leading to ankle work as the body sways back and forth, but the absorption and return are roughly equal, opposite and cyclic—the ankle behaving like a spring to hold the body upright. A more complex example is to view the leg as a whole, and observe that while the ankle generates power, the knee absorbs it; semi-active energy management can therefore transfer energy from one joint to another without having to generate it [29]. These cases allow the potential of mimicking the behavior with energy storing elements such as mechanical or pneumatic springs [30,31].

Other portions of biomechanical function can be viewed as accomplishing mainly a transition from one kinematic state of the body to another. For example, ankle angle changes in adapting to ramps or in standing on tiptoe are often made using active muscle work, but their final kinematic state may not need to change thereafter. Thus, a semi-active means of reorienting the ankle can mimic the overall function of adapting to terrains and tasks [20].

These and other ways of viewing joint behavior open a rich space of combinations and implementations to dynamically transition among multiple states and functions.

3 The semi-active design process

The process of developing semi-active devices that operate under these principles requires a challenging combination of theoretical analysis, mechatronic creativity, and pragmatic compromise. Starting from observations of human movement, the design team must overcome a series of hurdles to achieve a successful solution. These include a variety of steps in translating an observation into a semi-active device:

- notice a describe a specific biomechanical behavior,
- abstract it to a conceptual model,
- elaborate it to a mathematical model,
- explore the model's performance when the behavior is modulated,
- define measures to quantify the behavior of a human in terms of the model,
- compare the behavior among unimpaired persons and users of existing devices,
- conceive a semi-active technological realization of the conceptual model,
- reduce that technology to a design,
- transform the design into an artifact,
- control the artifact to mimic the behavior, and
- test the effects of the new system on humans, compared to the effects of other devices.

The following sections articulate these steps through a series of key questions that should be asked and answered through the development process.

3.1 Motivating a semi-active device

Before undertaking the design of a semi-active device—or any new device—it is important to have a well-justified motivation for doing so. This section outlines the process of developing this motivation based on biomechanical or biomimetic principles. The goal is to ensure that the resulting device addresses an important and impactful problem, set within a clear theoretical framework, in a way that can be compared and evaluated to prove its benefits relative to existing devices.

3.1.1 "What is the body doing?"—defining models of biomechanical behavior

3.1.1.1 Describing behavior

The first step in the process is to notice and describe a biomechanical behavior that might be valuable to emulate. This step sounds trivial, but it requires an important, learnable skill: the ability to look closely and critically enough to specify exactly what aspect of movement is of interest. An example would be the statement "The human ankle flexes a lot in walking, but very little in standing." This behavior can be observed in one's own body, or from video recordings, or if one happens to be comparing motion capture data of the two activities. Whatever the source, the insight is in the realization that the body modulates ankle behavior in a certain predictable way, and that this control may have consequences for movement.

It is important to distinguish between descriptions of "behavior" and descriptions of "function." For the purposes of this chapter, a rough definition will suffice: *function* is "what the body can do", whereas *behavior* is "what the body does." If the observation of interest is a *function*, the task of designing semi-active devices to mimic it becomes much harder: most of what the body can do is controlled by powerful muscles, and to replicate that capacity would require a powerful actuator. In the example stated, the *function* observed might be "the body can control ankle movement during walking and standing." This statement is correct, but to mimic it requires a fully active device. Focusing instead on observed behavior-what the body does in specific circumstances-narrows the scope of interest because most behaviors do not use the body's full functional capacity. In fact, the body performs most tasks as economically as it can, using strategies that minimize caloric expenditure and often mechanical work as well. In this sense, one could say the body tries to emulate a semi-active device in many tasks!

3.1.1.2 Conceptual modeling

To move forward with biomimetic device development, it is necessary to abstract the observed biomechanical behavior into a conceptual model that captures its importance to movement. This step is fundamentally a creative act, in which a high level movement is related to a specific behavior of interest in an integrative way. A conceptual model is not necessarily a unique description of the behavior; often there are multiple feasible concepts. But, a good conceptual model must capture cause and effect-it provides a means of articulating how, if the behavior of interest changes, it will affect the rest of the movement.

In the example of the ankle, one conceptual model could be articulated as a story: "in walking the ankle is loose and lets the body fall like a stick stood on end, but in standing the ankle is firm and holds the body upright like the base of a lamp." The words used in this qualitative concept-loose, firm-lead to the formal concept of the ankle as a spring with controllable stiffness: "the ankle behaves like a spring, with a high stiffness in standing and a low stiffness in walking." With the conceptual model of the ankle as a controlled spring, the mathematical tools of mechanics can be brought to bear in a formalized predictive model.

3.1.1.3 Mathematical modeling

The purpose of transforming a conceptual model into a mathematical model in the context of device design is to predict and evaluate how the device affects the movement of a person. Without a mathematical model, prediction is at the whim of intuition, and could be made differently by each person who considers the question. Different models can also make different predictions, and a given concept may lead to multiple mathematical models, but formalizing each model mathematically establishes a framework for organizing and relating the effects of the model's features and parameters.

For the case of biomechanical devices like prostheses, the mathematical formulation is usually based on the laws of statics, dynamics, and/or mechanics of materials. Which laws to incorporate depends on the type of behavior the device concept will mimic and the nature of its effects. For example, if the behavior is the various grips an opposable thumb can achieve, then a static analysis may be appropriate: it could show how forces are distributed to different fingers in different grips. Alternatively, if the behavior is the action of muscles in maintaining compressive loads on bones, then a strength-of-materials approach may be appropriate: it allows analysis of stress and strain in a loaded structure. In the context of movement and

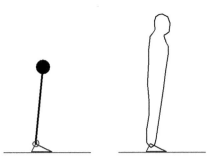

Figure 9.2 In standing balance (and in walking), the ankle is often modeled as a spring that increases ankle torque as the ankle angle deviates from neutral.

balance, the behavior is usually how control of a joint affects movement or stability-questions most appropriately answered through a dynamics–based model.

For the example of ankle control in walking and standing, a dynamic model is appropriate. The model needs to answer questions such as: How does the stiffness of a spring–like ankle affect the stability of an upright body in balance? How does the stiffness affect the speed and energy cost of walking? And how does the stiffness affect loading elsewhere in the body, such as contact forces on the foot and moment at the knee joint. These types of questions require not only an ankle model, but also a model of its interaction with other members. To study standing, the upright body is commonly modeled as an upright stick; this *inverted pendulum model* is among the most common biomechanical models of movement. To study ankle effects, the inverted pendulum body must also be coupled to a foot in contact with the ground, often modeled as a flat foot with specific heel and toe length. The spring-like ankle connects these two bodies, centering them in a neutral pose and applying torque between them if the ankle angle changes (Fig. 9.2). A dynamic model is built up from these component models, and possibly others such as another leg or a torso. The full dynamic model will include equations of motion for at least one body, kinematic equations and constraint equations defining the geometric interaction of bodies with each other and with the outside world, and constitutive laws for how forces and moments are applied to the bodies.

3.1.2 "What happens if the task is done differently?"—predicting the effects of device parameter changes

Once an appropriate model is available, the next step is to predict whether the biomechanical behavior captured in the model is likely to have effects on movement-beneficial or deleterious. How the model is used to make these predictions depends on the questions it is intended to answer. For semi-active device design, the intention is often to modulate the parameters of the constitutive models or the geometric constraints to change certain aspects of the system's movement. The conceptual question is therefore "How does the movement change if the parameters change?"

Mathematically, this question requires defining specific aspects of "the movement" that are of interest. For example, in standing balance, one question may be "Does the body remain stably upright?"; in this case the dynamic model is analyzed to show convergence or divergence of the system from the upright equilibrium, using stability analysis such as the eigenvalues of the linearized state-space system matrix. In the case of locomotion, one question may be "How much energy is dissipated when the next leg hits the ground?"; this question can be addressed by analytical or numerical propagation of the system state through the next leg's impact, and computation of the energy lost through impact equations [32]-[37]. Formal definition of such target quantities allows the model to predict specific benefits and costs of changing system properties.

These predicted effects can be used to understand the prospective value of a potential device designed to modulate specific properties. In closed-form models or model approximations, the sensitivity of some outcomes to the input parameters can be deduced analytically. Other models may be too complex or nonlinear, and simulations must be used. In any case, the results of the modeling process are predictions of how important movement outcomes-upright stability, energy loss, etc.-are affected by changes in the behavior originally conceptualized from observation of movement. These predictions provide a basis for deciding whether a semi-active device that modulates a specific parameter is ultimately worthwhile, or if instead a fixed parameter is expected to be adequate. For example, such modeling studies suggested that energy consumption should be strongly affected by leg push-off [35], foot shape [32], [33], [36] and ankle stiffness [37].

3.1.3 "How can we tell if the movement changes?"—developing metrics to quantify behavior

An additional important use of the model is to provide a perspective through which the movement behavior of a person can be interpreted. The modeling process started with a qualitative observation of behavior and resulted in quantitative predictions of its effects; the model can then be used to complement the original observation by providing a way to measure the behavior parametrically. This may sound like a simple intellectual exercise, but it is in fact critically important for ultimately validating a device designed through this process: without a quantitative definition of the original behavior, there is no way to determine how well a device mimics that behavior.

To enable such a comparison, the model parameters that grew from the conceptual model can be mapped back to measurable aspects of the original movement. In the example of the spring-like ankle, the spring's stiffness can be measured according to Hooke's law, which defines the behavior of a torsional spring with linear torque-angle characteristics: $T = k_{ang}\beta$ (for torque T, ankle angle β and angular ankle stiffness k_{ang}). Rearrangement yields a definition for the angular spring constant, $k_{ang} = T / \beta$. Thus, a simple metric to quantify the "loose" or "firm" ankle control initially observed is the estimated spring constant observed in human movement: measured ankle torque T divided by measured ankle angle β (or more generally, change in torque divided by change in angle, $\Delta T / \Delta \beta$). This is in fact a common metric in ankle biomechanics, termed the *ankle quasi-stiffness*, used both to measure human behavior and to specify biomimetic devices like prostheses [2], [31]. In a mechanical device, the ankle can be physically deflected by an applied moment and the resulting angle measured.

Not all behaviors are so straightforward to quantify based on the associated model parameters. For example, a more advanced dynamic model of the ankle includes not only stiffness but damping and effective mass as well (collectively *ankle impedance*), and requires dedicated experiments to measure [4], [38]-[40]. Nevertheless, it is critically important to the validation of a device to have a means to compare its performance to that of a human, compare a human's performance with versus without the device, and compare performance with different device settings.

3.1.4 "Do existing prostheses produce a deficit?"—evaluating the potential value

With the biomechanical groundwork in place–observations, conceptual and mathematical models, predictions of an effect, and metrics to measure performance and parameters–it is still not clear that a new semi-active device is needed. One final consideration is: just how badly do existing devices perform, and how much improvement can be expected from a semi-active version? What sorts of performance deficits to they exhibit, in their own function and in the resulting function and behavior of the user? How large and severe are these deficits, and how can it be established that a new device made an improvement?

The predicted effects of the conceptual device and the movement metrics defined in the previous step can be used to answer these questions. In many cases, an estimate can be made from clinical observations, prior literature, or prior data. Clinical observations sometimes align neatly with the concepts for a device. For example, every prosthetist knows that walking comfort and mechanics can be greatly influenced by even small changes in prosthetic component stiffness and alignment, and that these interact with ground slope in the environment [41]. This observation provides evidence for the potential value of semi-actively controlling ankle stiffness or angle. Prior studies in literature can sometimes support this qualitative understanding with quantitative evidence from past studies. For example, the concept of controlling prosthesis stiffness is related to past work quantifying ankle stiffness in unimpaired persons [2,31] as well as stiffness in prostheses [42-46] and the effects of stiffness changes on gait [28,47-53].

If prior knowledge in the field is inadequate to motivate the device concept under consideration, then a biomechanical study may be necessary to gauge the potential benefits of a new device. Undertaking a new study can be an expensive and time consuming proposition, but it is important to remember that comparative data will be necessary to show the ultimate benefit of the new device anyway. It is good practice to obtain and critically evaluate this kind of data prior to undertaking the development process, and doing so provides an opportunity to evaluate the new outcome metrics proposed in the previous step of the process outlined here. An alternative to a new study is to reanalyze prior data from past studies of the population under consideration (e.g., persons with amputation) or from variations in locomotion of unimpaired persons. The modern push toward data sharing in science may make such data sets increasingly available.

3.2 Designing the semi-active mechanism

Once proper motivation for the device concept is established, it is time to consider how to implement it in a semi-active design. How will the mechanism operate? If the concept is to modulate some property, how will the property be changed? How will it be maintained? What sensing will be necessary for the device to coordinate its operation? How will the device decide when to operate? How can the power required for these operations be minimized? How can fail-safety be built into the device? The following subsections introduce the key principles that address these questions, and the next section describes specific mechanisms that put these principles into action.

3.2.1 "How could a semi-active device accomplish the goal?"—design for semi-active actuation

Perhaps the most critical design challenge is conceiving a mechanism that can physically control the properties or behaviors the device is meant to modulate. The concept of a semi-active device eliminates high-power actuators from consideration, and places severe constraints on high force actuators as well. Almost any movement of an actuator during load-bearing periods requires high force, and movement through a useful range generally involves enough displacement that substantial work is performed. The semi-active goal of extreme low weight and power eliminates both these options. How, then, can a device be actuated?

The enabling characteristic of human movement is that it often subjects biomechanical devices to intermittent repetitive loading rather than continuous high forces. This is particularly true in lower-limb prostheses, which support loads well in excess of full body weight during stance phases of locomotion but are intermittently relieved of load during swing phases. During these unloaded periods, motion can be actuated with low force and high speed without requiring high power or substantial work. The same principle may apply to other biomechanical devices in certain circumstances, such as lower limb orthoses and hand prostheses (due to intermittent grasping). Furthermore, mechanisms may be designed to effectively off-load specific portions of a structure during different periods of a movement even if overall loading is continuous. For example, one well-known semi-active foot-ankle prosthesis-the CESR foot [30,54–56]–off-loads a latching mechanism whenever the foot force is applied at the toe, enabling the latch to operate even during stance phases. Thus, the key concept is to design the mechanism so that the actuators experience periods of minimal external force, and move them during those periods.

3.2.2 "How and when should the prosthesis adapt?"—Controlling semi-active adaptation

If the device is to move only when unloaded, it must be given the capacity to determine such a condition, and other conditions under which it should not move. Force sensors are a direct way to detect applied forces, but they can be inconvenient because they must be installed in series with either the device as a whole or with the property modulation mechanism. A common choice for initial designs is to use a force-sensitive resistor (e.g., [57]). However, such flexible devices can have problems with longevity and signal stability. A strain gauge-based design is better for long-term use, but it adds complexity, weight and cost.

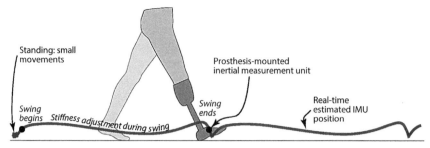

Figure 9.3 *Use of an inertial sensor to track foot motion and determine when and how to control a prosthesis.* Stance phases (*red*) (Gray in print version) have little motion, in both raw signals and reconstructed path. These features identify periods of likely high loading on the prosthesis, when the semi-active mechanism should not try to move. Swing phases (*green*) (Light gray in print version) have high-speed motion and low force, indicating a good opportunity for the semi-active mechanism to adjust the pros-thesis' properties. *Figure adapted from Ref. [58].*

A common alternative is to use an inertial sensor to detect movement, and infer that when the foot is not moving it is on the ground (Fig. 9.3). Inertial measurement units (IMU's) containing accelerometers, angular rate gyro-scopes and optionally magnetometers, are small, accurate and inexpensive. In many cases, stationary periods detected by low angular rate and accelerom-eter measurements do indicate contact with an external object, and hence the possibility of force application. Then, high signals suggest unloaded con-ditions during which the mechanism can be moved. Advanced movement reconstruction techniques [58-61] together with classification techniques can provide additional information to differentiate specific activities such as stair climbing, walking or running, to enable more advanced control.

Unfortunately, motion-based contact detection can sometimes be fooled. For example, stationary periods with a foot or limb elevated can lead to false contact detection (a missed opportunity to adapt the mechanism). Or, activities with dynamic contact can cause false detection of unloaded phases–a potentially harmful condition causing the device to try to adapt when it in fact cannot move. To prevent potential damage in these circum-stances, motion-based contact detection should be combined with a sec-ondary sensor to verify mechanism movement and shut down the actuators if they are in fact unable to move.

3.3 Evaluating performance in semi-active devices

Finally, any device developed to improve health, function or mobility must ultimately be evaluated for its effects. Some of the steps outlined earlier in the design process were included specifically to enable this evaluation phase: defining a model, deriving outcome metrics from it, and evaluating shortcomings in existing devices' performance. These steps provide ready means of comparing the new device's effects against the prior state-of-the-art.

3.3.1 "Does it work?"

Validation of the mechanism itself is the most straightforward proof of success in semi-active design. This is a simple test of whether the system achieves the basic function it was created to achieve. For the example of a controllable ankle stiffness, the stiffness of the prosthesis can be quantified through mechanical testing in a variety of settings, and the range compared to that specified based on the natural ankle in preliminary work.

Several additional performance questions are also relevant to semi-active devices. How long does it take to effect a change of parameters? Too slow and the device will be unresponsive; too fast and it may have been overbuilt (too heavy and powerful). How well does the device maintain its parameter settings? A device that can maintain settings without continually powering its actuators will be more efficient than one that must power them. And, how precisely are the parameters controlled? Are there ways of optimizing the control to improve precision while minimizing power, or to find the best compromise between these goals? Finally, do the size, weight and power of the device fit within the allotted limits? Will it last long enough on a small enough battery to be unobtrusive and convenient? These are some of the many important outcomes that must be achieved to have a successful semi-active device, and may be among the specifications reported as part of its description.

3.3.2 "Does it have the intended effect?"

The strongest establishment of benefits is a head-to-head comparison of the new system against existing devices. This style of study is used to investigate the null hypothesis that the effects of the new device do not differ from those of one or more existing devices. If statistical comparison of the relevant outcome metrics across these devices rejects this null hypothesis, then the benefits of the new device can be established. In this type of comparison, the hypothesized benefits should include a "signed" benefit: greater values for a desirable outcome or lesser values for an undesirable one. For example, the well-known semi-active CESR foot-ankle prosthesis was designed to use semi-active latching mechanisms to recycle energy from heel strike to increase ankle push-off work [30,54-56]. In cases like this, a direct comparison against an existing device can establish a clear benefit.

More typically, a semi–active device is developed not to overcome an absolute deficit in some quantitative metric, but rather to control properties to produce multiple different behaviors—that is, to establish adaptability. In such cases, determining a proper comparison against existing devices is not always straightforward. An alternative approach is to investigate the new semi–active system on its own, to show how well it demonstrates the intended modulation of properties and performance. For example, a recent semi–active variable-stiffness prosthesis was evaluated for its ability to control the amount of energy stored and returned during walking [58]. This evaluation is done to show the proper and intended function of the device itself, without a direct comparison to establish superiority to other devices. It is still helpful to compare performance against existing devices based on literature or prior research, to demonstrate the new device's range of adaptability compared to the state-of-the-art.

The practical benefits of this adaptability-as opposed to theoretical benefits-can be challenging to demonstrate, so it remains worthwhile to design a well-structured comparison against existing systems. Particularly with devices designed for adaptability, the user must be tested in a variety of circumstances chosen to elicit adaptation in the device and thereby demonstrate the contrast with existing devices. Examples may include standing versus walking, walking on different slopes or terrains, walking versus running, or walking and running across a range of speeds-any circumstances in which adaptability is important. Some of these comparisons may involve non-traditional means of data collection such as field-based tests with wearable sensors [61-64] or portable motion capture [65,66].

3.3.3 "Do people like it?"

The model-derived metrics developed to evaluate biomechanical behavior and compare across devices fall under the category of biomechanical benefits: quantitative measurements with theoretical benefits that can be used to argue superiority. But ultimately it is the user's experience with a device that will determine its acceptance in the field. Therefore it can be helpful to include a functional capability test or questionnaire and/or a preference survey among the outcomes of a study, to gauge whether the new device improves this user experience. A variety of validated tools exist for purposes including gauging mobility capacity using functional tests and gauging real-world functional mobility based on users' experiences and confidence. Table 9.1 lists a sampling of general-purpose and amputation-specific tests and questionnaires that may be useful for gathering subjective feedback. Prior to

Table 9.1 Sampling of general-purpose and amputation-specific tests and questionnaires.

Instrument	Format/purpose
Amputee mobility predictor [67]	Battery of physical mobility tests to gauge mobility and rate an individual's "K" level.
Prosthetic Limb Users Survey of Mobility (PLUS-M) [68]	12-item questionnaire (short form) to gauge how well a prosthesis user can accomplish locomotor tasks such as walking on level and uneven terrain, responding to perturbations, carrying things, and hiking.
Activities-Specific Balance Confidence scale (ABC) [69]	16-item questionnaire to gauge individuals' confidence that they will not fall in various balance-perturbing circumstances such as reaching, bending, climbing stairs, being jostled, or entering/exiting a car.
Medical Outcomes Study- (MOS) Short Form 36 [70]	36-item questionnaire to gauge general health. Subsections relevant to movement include vitality, physical functioning, bodily pain and physical role functioning.
NIH PROMIS (short forms 10, 11, 20) [71]	10, 11, or 20-item questionnaire to gauge how much individuals' health limits their activities including self-care, work and locomotion.
Locomotor Capabilities Index (LCI-5) [72]	14-item questionnaire to gauge how well a prosthesis user can locomote, including items like getting up after a fall, walking in bad weather and climbing stairs with versus without a handrail.
Prosthetic Profile of the Amputee (PPA) [73], [74]	44-item questionnaire to "evaluate and to determine the factors potentially related to prosthetic use by a person with a lower extremity amputation after discharge from rehabilitation" [74].
Prosthetic Evaluation Questionnaire— Mobility Subsection (PEQ-MS) [75]	12-item self-report questionnaire for evaluating mobility, including walking on level, unlevel, and slippery surfaces and getting into and out of chairs.
Prosthetic Mobility Questionnaire [76]	12-item self-report questionnaire for evaluating mobility, similar to PEQ-MS (above) but with some different items including vigorous activities (running, long walks).
Socket Comfort Score [77]	One-item, 11-point scale to evaluate the comfort of a prosthetic socket in a clinical environment.

completing such instruments, research participants testing a semi-active device should attempt a variety of tasks spanning different functional challenges, so that the costs and benefits of the device's adaptability can become clear to them. Free-form commentary from users can also be highly beneficial as feedback and motivation for the device developers.

3.4 Iterate, iterate, iterate

The final step in the design process is to remember that the process is always ongoing-no device is perfect and all can be improved. The results of any stage outlined above can be turned back into new insights for other stages in the design and testing process. Testing and user feedback are particularly valuable for motivating future generations of semi-active devices, as the function of these devices must be specified relatively early in the design process, compared to fully active robotic prostheses where functional changes can be mainly a matter of control. Findings that alter this relationship can inspire new or different versions of a semi-active prosthesis to improve its benefits to the user.

4 Design principles for semi-active mechanisms

The design process laid out above is relevant to any biomimetic device, not only semi-active systems. So, what is it that distinguishes the design of semi-active devices? The key idea in a semi-active mechanism is to exploit opportunities to change a device's functional parameters when the change itself requires little force or energy. The goals of this approach include minimizing the size, weight, cost, and power consumption of the actuators and the mechanism. This section outlines principles that can be exploited in semi-active prostheses to enable this low-power, lightweight regulation of properties.

4.1 Exploiting the gait cycle

Perhaps the most important idea in semi-active prostheses for the lower limb is to recognize the opportunity provided by the cyclic loaded and unloaded (stance and swing) phases the prosthesis experiences during gait (Fig. 9.4). During stance phases the foot supports forces in excess of body weight (around 1000 N), whereas in swing phases it supports no load at all. Stance phases are when the prosthesis' properties influence the biomechanics of the legs and body, and swing phases provide an opportunity to change these properties without having to overcome large external loads.

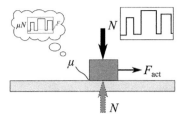

Figure 9.4 *Conceptual illustration of how intermittent applied external force N leads to periods of very high and very low friction force μN to resist the motion of a mechanism.* Semi-active actuation can be applied when the friction force is low—typically when the foot is off the ground. Similar intermittent forces may apply to different parts of a mechanism at different points in the gait cycle depending on device design.

The periodic application of body weight forces can also be a source of actuation: it can provide energy input to the mechanism from movements elsewhere in the body, and can sometimes allow certain actuators to be eliminated from the design. Each of the mechanical design principles below exploits this cyclic loading and movement to improve or enable its semi-active operation.

4.2 Clutches and latches

One key principle is the use of clutches and latches. Both these elements have two modes: a free-wheeling mode in which they allow relative movement between two parts, and a locked mode that prevents such movement. A clutch is capable of locking in many positions throughout the motion and is often a continuous, friction-based mechanism, whereas a latch usually operates in a single position and often involves physical blocking of the movement with a distinct component. Both mechanisms are useful for semi-active mechanisms because their action does not inherently require the addition or removal of energy from the machine. Clutches and latches are frequently employed to allow movement in one direction but not another, or to hold and release large loads-for example, in energy storage elements during different phases of a movement. The ideal tool for this purpose is a controlled clutch or latch that can turn on and off at any time, but the design of controlled clutches that support body-scale loads is a field of active research.

A good example of using clutches and latches to save power is the Controlled Energy Storage and Return (CESR) foot prosthesis [30] (Fig. 9.5). The CESR foot was designed to perform energy recycling by using a one-way clutch to allow storage of energy in a spring but prevent its release. In

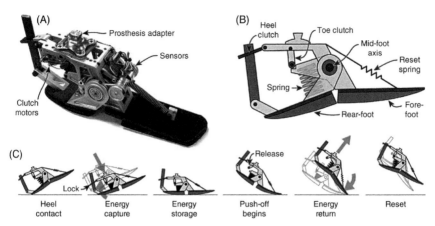

Figure 9.5 *The Controlled Energy Storage and Return (CESR) prosthesis uses a clutch and a latch to capture energy in a spring at each heel strike, then returns the energy semi-actively, using lightweight motors to release the latch and reset the mechanism. Figure reprinted from Ref. [30].*

this prosthesis the hindfoot and forefoot members are independently articulated, with a spring between them. Upon heel strike, body weight moves the hindfoot member and compresses the spring to store energy, while a linear clutch like that in a bar clamp displaces to allow the motion. When weight transfers to the forefoot, the clutch uses wedging action to hold the bar and prevent recoil of the spring. Later a latch is released from the forefoot side of the spring to allow energy release through forefoot push-off. Finally, the clutch is released to allow the mechanism to reset. The clutch and latch in the CESR mechanism are very efficient and low-power: electrical energy is used only to release the latch and the clutch with two small motors after bodyweight loads are removed from the parts they restrain. Thus, the CESR successfully cycles roughly 20 W of mechanical power through the spring [55] while consuming less than 1 W of electrical energy [30].

4.3 Non-backdriveability

Another key idea in semi-active prostheses is designing systems that are inherently non-backdrivable. In a non-backdrivable system the actuators can move the mechanism, but external forces cannot. In a backdrivable system external forces are transmitted through the mechanism all the way back to the actuator, which must actively counteract them to prevent undesired motion, or alternatively incorporate a parallel clutch, latch or brake to hold the load. Non-backdriveability removes this demand from the actuators: the small actuators are not capable of driving the mechanism through large external forces, and the large forces cannot force the mechanism to move either, so the actuators can simply be shut off when external forces are present. Such intermittent shut-downs are a very effective way to reduce power consumption and control complexity. This approach can be thought of as "set it and forget it" control of the device's properties.

4.3.1 Friction as an ally

Non-backdriveability is usually a consequence of friction in the mechanism. For example, high-ratio gear drives are often non-backdrivable due to frictional losses at each stage of the transmission. In powered systems, one definition of non-backdriveability is the loss of more than 50% of the input power during actuated movement. The same definition can apply to semi-active systems, but since the power input is small, so is the power loss.

4.3.2 Orthogonal and Inclined-plane actuation

A simple machine that is often non-backdriveable due to friction is the inclined plane, commonly implemented as a worm gear or power screw in which the inclined surface drives a movable object (Fig. 9.6). In these examples, if the friction coefficient μ is greater than the tangent of the incline angle γ, the system cannot be backdriven by a vertical contact force. Because friction is notoriously variable in practice, systems designed to be non-backdrivable are typically designed with a contact angle well below this threshold, limiting this approach to inclines of relatively shallow angle. For example, with a friction coefficient of 0.1 the theoretical allowable incline angle is only 0.1 radians or 5.7 degrees, so the designed incline should be lower than this threshold. However, sometimes additional tricks

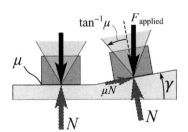

Figure 9.6 Illustration of the friction-lock concept on an level surface and an inclined plane. External loads will not backdrive the mechanism (*shaded square*) as long as the applied force $F_{applied}$ is within the friction cone, $\gamma \leq \tan^{-1}\mu$. The available friction force (limited to μN by friction and the normal force) will resist the down-slope component of the applied force.

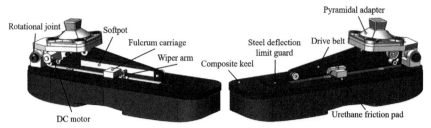

Figure 9.7 *Variable Stiffness Foot (VSF) that controls its forefoot stiffness to vary energy storage and return.* The critical feature is the fulcrum carriage that supports the keel: moving the fulcrum toward the forefoot increases the forefoot stiffness; moving it toward the heel decreases stiffness. Another enabling feature is the urethane friction pad which prevents the keel from backdriving the carriage under high deflections. The keel itself provides the stabilizing reaction force through friction with the pad. This feature enables "set-it-and-forget-it" control and minimizes the necessary actuation.

can increase this non-backdrivable angle, such as using friction on multiple contact faces (as in the TADA prosthesis below).

For the limiting case where the tendency of an incline to back-drive is perfectly balanced by friction, exactly 50% of the work input to drive the inclined plane is lost to friction. A powered prosthesis would not be designed with such a high mechanical loss, but a semi-active prosthesis can be built to exploit it. The key once again is the cyclical nature of gait: when the foot is off the ground, the load on the prosthesis' mechanism–and therefore energy needed to move it–is low enough that a 50% loss does not matter. Then during loaded periods, non-backdriveability allows the actuator to be turned off while the mechanism stays in place. Friction is used in this way to improve the non-backdrivability and lower the power consumption of the Variable-Stiffness Foot (Fig. 9.7, 9.18).

This set-it-and-forget-it control should not be pushed too far in practice. In the same way that machine screws can vibrate loose over time, vibrations and loading/unloading cycles will gradually cause a friction-locked mechanism to drift away from its intended setting. Therefore the "forget it" part should be used sparingly: it is still wise for the system's controller to check for drift and correct it periodically, perhaps even every stride. On this time scale the errors are small and the corrections quick, so little power is consumed to maintain a setting.

4.3.3 Kinematic singularities

Another mechanism that can provide non-backdriveability is a linkage with one or more kinematic singularities. Linkages that control a continuous system output–such as robot arms–are generally designed to avoid operating near kinematic singularities in order to have well-conditioned mathematical forms for computing the control variables. But, kinematic singularities can be beneficial in some circumstances, such as when people lock their knees during standing: the slightly hyperextended posture of the knee joints allows ligaments to support this posture rather than muscles, saving energy and eliminating the need for neural control at that joint. This action effectively eliminates a degree of freedom from the leg linkage, and thereby makes the linkage more stable.

Kinematic singularities can be exploited in a mechanical linkage to eliminate actuator loads in certain configurations. Much like the human knee, a linkage operating away from its joint singularities requires actuators to control those joints, often with a mechanical disadvantage relative to the external loads. But when a joint is straight (or fully collapsed–see the Rock'n'Lock Foot, Figs. 9.8, 9.12, 9.13), external loads create little or no joint torque and therefore little or no actuation is required.

Singularities can also be used to lock the mechanism to support high loads. If the mechanism is a four-bar linkage, placing one joint into a singularity also removes the mechanism's last degree of freedom. Then the

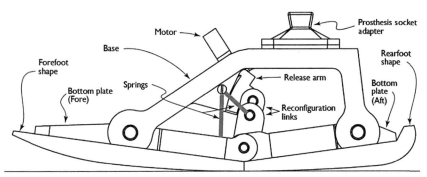

Figure 9.8 *Mechanism of the Rock'n'Lock bimodal prosthesis in walking ("Rock") mode.* The key feature is the two-link controllable singularity mechanism consisting of the upper/lower locking links and the release arm (or "nudger"). Hidden at the lower joint is a bushing for the lower axis that allows sliding relative to the hindfoot link, to prevent an overconstraint condition. The upper spring acts to pull the joint between the two links lightly toward the forward hard-stop, in both modes. The lower spring stores energy to lift the midfoot and collapse the linkage in the transition from walking mode to standing mode. The servo motor on the top surface has its axis orthogonal to the plane of the mechanism, intersecting the upper axis of the linkage. An arm rotates 180 degrees about this axis, pulling on the upper end of the "nudger" with a preloaded spring to initiate mode transitions.

linkage becomes a truss, which can rigidly hold its shape against very high external loads (Fig. 9.13). Switching in and out of different singularities can thus enable a linkage to behave like different mechanisms at different times.

It is important to note that singularities can be either stable or unstable, and the difference affects how they must me managed. A stable singularity needs no management at all. For example, the elbow joint is in a stable singularity when a person hangs from a pull-up bar with arms straight: if the elbow is perturbed from its straight configuration, tensile forces along the arm will pull it straight again, back into the singularity. By contrast, the knees in standing are in an unstable equilibrium: a common children's prank is to poke a person's knees into slight flexion, causing the legs to collapse. Unstable equilibria need special treatment-in this case an over-center mechanism biased in one direction (slight knee hyperextension) with a hard stop to prevent collapse in that direction (the posterior ligaments of the knee). Often springs can be configured to lightly bias the mechanism into the hard stop (e.g., Fig. 9.8).

4.4 Springs to replace actuators and simplify control

Springs can play an important role in semi-active prostheses, as energy storage elements that can drive certain motions without the need for controlled actuators. Of course, springs cannot provide net positive work, but they provide convenient energy storage elements that can harvest energy at certain times and return it to actuate a mechanism at other times. This function is particularly useful in prosthetic feet due to the cyclic loading of the gait cycle: in every stride, the body applies large loads that can drive energy into a spring, which can then be latched, clutched, and otherwise managed to release the energy in beneficial ways. Springs also function as convenient always-on force sources that enable a system to be mechanically responsive to changes in state without high-bandwidth control. In this way, springs are a critical component enabling some forms of the set-it-and-forget-it control described above.

The CESR foot described above is a prime example of using springs to replace multiple actuators and simplify control. The central spring is the main energy storage element; it is compressed by body weight upon loading of the heel. But, there are three other important springs in the mechanism as well. One is a spring that holds the heel clutch into a locked configuration during load-up of the heel; without this spring, the clutch would not lock and the whole mechanism would not function. This spring is opposed by a one-way motor, spool and cable actuator that briefly releases the clutch to enable the final reset of the mechanism. Another secondary spring holds the toe clutch in position to prevent the forefoot from plantarflexing when the main spring is loaded. This spring is also released by similar cable mechanism. Both these clutch springs store small amounts of energy but enable their opposing actuators to be active for very short periods of time; the springs then apply a continual restoring force that waits without consuming energy until the state of the mechanism allows them to move. The last spring in the CESR is the reset spring, which is stretched as the forefoot plantarflexes under the power of the main energy storage spring. This arrangement is unusual because it uses the recoil of a strong spring to drive the loading of a weaker one, which then performs a separate function (reset) at a later time. Thus, body weight loads the main spring, the main spring loads the reset spring, and finally the reset spring moves the mechanism back to its neutral position after the heel clutch is released.

Another use of springs is to insulate internal switching components from potentially-damaging high external loads. The heel clutch spring in the CESR provides an example of this function. The locking plate of the clutch must be held with a downward force to allow the clutch bar to slide through it, but the plate also articulates with the bar slightly to move with the kinematics of the linkage. If an actuator were used to regulate this movement, it would require precise high-bandwidth control, and there would be a high probability of backdriving the actuator whenever an irregular step occurs. But the spring in combination with the cable drive effectively insulates the actuator: if the clutch plate moves up too quickly, the spring extends with it and the cable goes slack. Thus, a tiny motor with one-way actuation is able to survive in its role of regulating body-weight forces through a clutch. For additional examples of how springs can insulate actuators from high loads, see the Rock'n'Lock foot example (Fig. 9.8).

The main challenge of using springs to replace actuators is to balance the springs and actuators that control the movement. In the CESR case above, the main spring is very strong and the reset spring very weak; in this case, they are dramatically imbalanced and the challenge is minimal. In the Rock'n'Lock foot below, three springs interact to control the tipping point of a bi-stable mechanism. In such a case, the spring properties must be balanced, and this can bring challenges for tuning. Therefore, it is necessary to pay close attention to the material, geometric, and fatigue properties of the springs, and to protect them from wear or damage, in order to have a reliable and durable mechanism.

4.5 High static forces, quick unloaded movements

The concept of switching during swing phases and holding during stance phases affects not only the power required from the actuator, but also the forces experienced at the joints. In powered systems, actuators are used to move the mechanism even when large external loads are applied, and this behavior compels the use of traditional ball and roller bearings in the joints, to minimize frictional losses and wear on the joint surfaces. Semi-active systems experience the same large forces but do not try to move when they are applied. Therefore, the risk of wear in the joints due to motion under high load is eliminated.

This combination of high static forces and no motion leads to distinct requirements for bearing surfaces in the joints of the mechanism. Traditional roller and ball bearings actually perform quite poorly under static conditions and have relatively low static load ratings. In a semi-active system, they may never experience the high-load, high-speed movements in which they perform best. In contrast, simple bushings (also called plain bearings) made of materials such as reinforced PEEK (polyetheretherketone) plastic have very high static load ratings for their size, and in a semi-active system they will never experience the high-load, high-speed movements in which they perform poorly. Such bushings are also much smaller, lighter, simpler and cheaper than traditional bearings, making them an ideal solution for rotational and sliding joints in prostheses, where weight is an important consideration. All of the example semi-active prostheses described below make use of bushing-type interfaces to exploit these benefits.

4.6 Simplicity and durability

Finally, it is important to emphasize the value of simplicity in semi-active prostheses. It is not the purpose of semi-active design to replace a straightforward actuated system with a semi-active system that is impossibly complex and difficult to manage. Prostheses need to last for several years and millions of cycles, in a wide variety of environmental conditions, all at a modest material cost and a reasonable market price. Therefore, simpler is better, and the potential for preventive and responsive maintenance and extended service life should be considered at the concept development phase. Such considerations may include the ability to easily tighten joints, replace consumable components, clean the mechanism, dry it when it gets wet, and otherwise freshen its operation with minimal professional intervention. It is also important to consider the impact of parameter variations, such as in friction or motor performance, to ensure operation with no maintenance for long periods of time. The field of semi-active prostheses is young and few designs have been subjected to rigorous testing, but as new systems with improved functionality move from concept to product, these real-life challenges must be addressed.

5 Example concepts and design implementations

Building on the processes of bio-inspired design and the key concepts for semi-active design, this section outlines the designs of several semi-active foot-ankle prostheses developed after different concepts of foot-ankle behavior. Each prosthesis is intended to mimic a different aspect of foot-ankle function or approach that function from a different perspective. Each prosthesis is described according to its underlying biomechanical concept, and the mechanism design and performance of each is described along with its relative strengths and weaknesses.

5.1 The foot as a wheel

5.1.1 Description of behavior: rolling

The first conceptual model of foot/ankle function is a wheel. The idea is that during the stance phase of walking, the action of the body moving forward over the leg looks very much like rolling on the rim of a wheel (Fig. 9.9). This "rimless wheel" view of leg function in walking has been active in

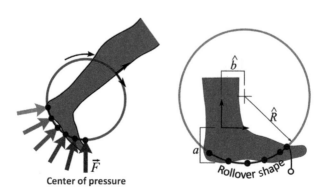

Figure 9.9 *Model of human ankle behavior as the rolling of a wheel.* (*Left*) During walking, the center of pressure under the foot advances as the shank angle advances. During single stance, the COP traces out an arc in a shank-based reference frame. (*Right*) This *Rollover Shape* can be parameterized with a radius R and forward offset b relative to the ankle (shown at height a above the ground). Rollover shape is a popular way of summarizing natural ankle mechanics and comparing prosthetic feet.

prosthetics since the 1800s [78] and has been a component of many simulation models of bipedal locomotion, for example, [32,36,79], but analysis of its detailed parameters and effects began in the 1990s [80]. The rolling concept captures the advancement of the center of pressure (COP) of ground reaction forces under the foot as the leg rotates forward. In the context of semi-active prostheses, the concept is to influence gait and balance by modifying the geometry of the rolling shape.

5.1.2 Mathematical model: radius of a wheel

A mathematical model of the relationship between COP and leg angle can be derived from the mechanics of a circular rolling wheel. In a rolling wheel, the COP is always beneath the center of the wheel and the differential motion of the COP (Δ_{COP}) relates to the differential rotation of the wheel (Δ_θ) according to the wheel radius R: $\Delta_{COP} = R\ \Delta_\theta$. This differential form can be used to estimate an instantaneous effective radius of the foot-ankle system from measured differential changes in leg angle and COP [81]. Alternatively, a summary measure of rollover radius can be made by tracing out the path of the COP in a shank-based reference frame to find the commonly-used "rollover shape", and fitting a circle to the resulting path to find its radius and forward offset [80], [82]. The rollover shape has been used to characterize normal gait, to compare different prostheses, and as a guide for prosthesis alignment [83]. Healthy rollover radius has been found to be about 0.3 times leg length in walking, and is conserved across changes in walking speed, carried load and heel height [82,84,85].

5.1.3 Variations: effects of different radii

The effective radius used by the healthy human leg is constant, but modeling studies indicate it has potentially important effects on locomotion if it can be changed. At the lower limit, the foot has zero radius and becomes a point, leading to jarring impacts as the body falls onto each successive support leg [34,36]; in the special case where the foot has radius equal to leg length, the two legs behave like a synthetic wheel [36,86]: the system rolls with a level, horizontal hip joint trajectory on each leg, and alternates legs without energy loss at the transition (Fig. 9.10). Between these special cases, continuous variation of foot radius gradually reduces the energy loss as the feet become more wheel-like [32,36], an effect ultimately attributed to the increasing length between the foot's initial and final contact points [33].

This model-based finding leads to the hypothesis that energy use in walking can be reduced simply by increasing the radius and length of an

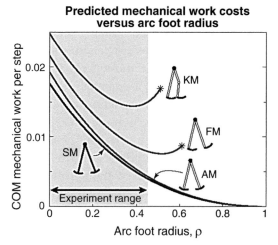

Figure 9.10 *Walking models with arc-shaped rolling feet demonstrate a strong relationship between foot size (here correlated with radius) and mechanical energy required to walk.* These model findings suggested a rolling foot-shape intervention in prosthetics. *Figure reprinted from Ref. [33].*

arc-shaped foot-an approach quite unlike the usual natural strategy of using ankle push-off to power gait. This hypothesis can be tested in humans by physically locking the ankle joint with an external brace and then changing the shape of a rigid arc-shaped sole attached below the brace (Fig. 9.11). Using this manipulation experimentally, humans do indeed show reduced mechanical energy use in the legs as foot radius and length increase [32,33].

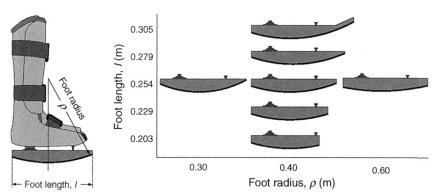

Figure 9.11 *Arc-shaped interchangeable foot bottoms demonstrated the effects of foot length and curvature on the work and metabolic energy requirements of walking.* Humanlike foot length is the most economical (0.29 leg length), and curvature has little effect. *Figure adapted from Ref. [33].*

In fact, at an intermediate arc foot radius slightly below the humanlike rollover radius, the mechanical work of walking actually falls below that of natural walking. And, at a foot radius and length roughly equal to natural the metabolic energy use is also minimized [32,33]. Thus, the conceptual model of the leg as a wheel leads to a potential means of saving energy in walking: simply by increasing the radius of an arc-shaped foot, an approach quite unlike the usual natural strategy of using ankle push-off to power gait.

The hypothesized benefits targeted walking, but that is not the only activity a person undertakes while upright on their feet. Individuals using such arc feet rapidly discover a problem with implementing them in a practical system: with small- to medium-radius arc feet, it is impossible to stand still! The problem is due to the bracing of the ankles: without ankle control, the body cannot control the center of pressure under the feet, and therefore behaves like a tower tottering on a rounded base. In this case the only way to stay upright is to continually move the feet to new locations to perform active balance. The problem can be corrected with a base that has large radius or a flat or concave shape; such a flat foot shape achieves stability [87], but leads to discomfort and increased energy expenditure in walking. Thus, the conditions necessary for walking and standing are at odds-a perfect opportunity for a semi-active device to enable both cases at different times.

5.1.4 Implementation: the Rock'n'Lock foot

The *Rock'n'Lock* foot is designed to provide the benefits of a rolling foot in a foot-ankle prosthesis used for walking, together with the benefits of a flat foot in a prosthesis used for standing balance (Fig. 9.12, 9.13) [21]. The materials and structure were chosen to match the properties of the equipment used in the underlying experiments: rigid shapes with no appreciable energy storage or return, rounded for walking and flat for standing. The approach is to incorporate two articulating segments of the prosthesis' plantar surface-a forefoot segment and a hindfoot segment. In walking the toe and heel lift up and the midfoot extends downward to provide a continuous rolling surface (shaped in EVA foam), whereas in standing the midfoot lifts up and the heel and toe extend downward to provide stable, heel-toe contact.

5.1.4.1 Semi-active design features of the Rock'n'Lock

The Rock'n'Lock incorporates several of the key features of semi-active device design outlined above. First, it uses latching mechanisms to lock the moving parts into multiple modes without continuous use of an actuator. In this case the latching mechanism consists of two planar links that lock into

Figure 9.12 *The Rock'n'Lock foot, in* (Top) *convex "Rock" mode for walking and* (Bottom) *concave "Lock" mode for standing.* The internal mechanism that achieves this switch is described in Fig. 9.13.

two kinematic singularities (Fig. 9.8, 9.12): one with the links extended to hold the midfoot in its downward position (rounded walking mode), and the other with the links nested to hold the midfoot in its upward position (concave standing mode). In both cases the middle joint between the two links is aligned a fraction of a millimeter anterior to the line between the

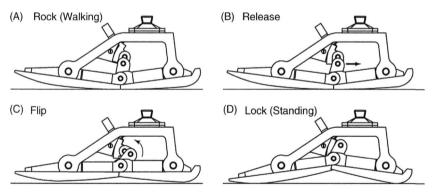

Figure 9.13 *Mode switching in the Rock'n'Lock foot.* In walking mode (A), the two internal links are in an extended configuration, with their common joint just anterior to its kinematic singularity, held forward by a spring (not shown here; see Fig. 9.8). Forces at the heel or toe tend to extend this linkage in tension, whereas forces at the midfoot tend to collapse it, forcing the joint against the hard stop. To switch to standing mode, the release arm (the "nudger") inside the mechanism nudges the joint out of its singularity (B), allowing the links to collapse toward the posterior (C) and nest together. In standing mode (D) the links are in a collapsed configuration, with their common joint again anterior to this second kinematic singularity, again held by the same spring (Fig. 9.8). Here forces at the heel or toe force the joint against the hard stop, whereas forces at the midfoot tend to push the joint upward into the stable singularity.

upper axis (connecting the first link to the core of the prosthesis) and the lower axis (connecting the lower link to the forefoot and hindfoot segments). The middle joint is pulled toward the front of the foot by a lightweight spring, to hold it against structural stops built into the base structure and adjusted with screws. Both singular configurations lock the whole system against both tensile and compressive forces transmitted from the lower axis: when these forces place the upper link in compression, the linkage tends to buckle at the middle joint, but no motion occurs because it is supported by the structural stops; when the upper link is in tension the middle joint is pulled or pushed away from the stops into a stable tensile configuration, but is pulled back against the stops by the spring when the load is removed.

Second, the Rock'n'Lock switches modes using minimally actuated mechanisms based on a low-power motor, springs and intermittent application of body weight. Two rotational axes allow the forefoot and hindfoot segments to rotate 15 degrees in the sagittal plane at locations approximating the metatarsal heads and the center of the calcaneus in the natural foot. The two segments are joined to each other and to the lower mid-foot link at the lower axis of the locking mechanism. This linkage would be rigid with only pin joints, so the hindfoot segment contains a slot with a PEEK slider bushing on the axle to allow relative motion. The foot segments are drawn into the concave standing configuration by a second spring pulling upward on the hindfoot segment. This spring cannot push the middle joint out from the extended configuration singularity (walking mode), but once the control mechanism pushes it out, the spring raises the whole linkage into standing mode. The switch to walking mode is actuated when the switching mechanism pushes the middle joint of the linkage out of the collapsed (standing mode) singularity and the user steps on the heel or toe. Body weight forces overcome the retraction spring, the locking linkage flips into its extended singularity, and the forefoot and hindfoot segments attain the walking configuration.

Third, the Rock'n'Lock uses set-it-and-forget-it control that allows the mode-switching mechanism to operate even when the prosthesis is loaded and cannot move, e.g. when the user is standing or has his/her foot on the ground during walking. The central feature of the control mechanism is a release arm called "the nudger" mounted on the upper axis of the locking mechanism. The nudger rotates through a small angle to allow the linkage into one singularity or the other, but not both. When the upper end of the nudger rests against its stop, the lower end protrudes to prevent the linkage from locking in the extended singularity, or to release it if it is currently locked. Conversely, when the lower end of the nudger rests against its stop, the upper end protrudes to prevent locking in the retracted configuration or to release it if locked. To switch modes, the motion of the nudger is driven through a spring-loaded tension link by a small servomotor. The spring connection allows body-weight forces to back-drive the linkage without damaging the actuator, but returns the nudger to its intended configuration whenever body weight is removed. The servomotor is mounted with its axis orthogonal to the upper linkage axis, intersecting it, and positioned along the line that bisects the two endpoints of the motion of the nudger's spring attachment point. The orthogonal axis allows the spring to be pre-tensioned, as it is never fully relieved of tension by the movement of the motor or the nudger; this allows the force it applies to the nudger to release the linkage to be higher than the motor could normally apply. The 180-degree rotation of the motor arm allows it to be deactivated once it reaches its set position, as the spring does not apply appreciable torque about the motor axis to backdrive it.

These design choices converge to allow the Rock'n'Lock to use very little power and therefore minimize overall weight (830 g). The motor operates only once each time the mechanism switches modes, which is once every time the user starts walking and once when the user stops. These behaviors are easily detectable by an onboard inertial sensor; the Rock'n'Lock as built detects movement using a single-axis angular rate gyroscope detecting sagittal plane angular velocity, together with a moving-average filter and thresholds for stand-to-walk and walk-to-stand transitions. Additional energy savings could potentially be realized using lower-power sensors such as accelerometers only. Whenever the user continues with either walking or stationary activities, the only energy used is for the sensors and the microcontroller that computes the decisions.

5.1.4.2 Performance and benefits of the Rock'n'Lock

To determine the functionality of the initial Rock'n'Lock prototype, it was tested by four persons with unilateral transtibial amputation, in both walking and standing tasks. The standing task consisted of standard posturography tests on two force plates: the subjects stood still for 1 minute in normal parallel-foot stance, using their customary prostheses and each mode of the Rock'n'Lock. The walking task consisted of level treadmill walking at 1.25 m/s. For standing, we measured center of pressure fluctuations of the whole body and weight support on each leg. For walking, we measured metabolic energy expenditure using indirect calorimetry (Fig. 9.13 Figure 9.13).

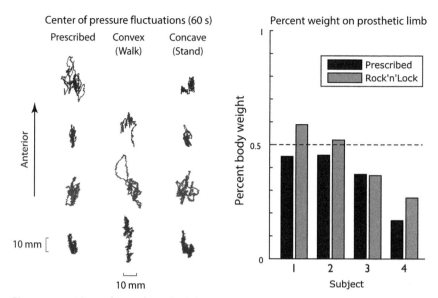

Figure 9.14 *Biomechanical results bilateral quiet standing with the Rock'n'Lock foot compared to four subjects' prescribed ESR prostheses.* (*Left*) Center of pressure fluctuations in the anteroposterior direction appear increased in the convex walking mode of the Rock'n'Lock, but similar or decreased in the concave standing mode. (*Right*) The fraction of body weight supported by the prosthetic limb is increased in Rock'n'Lock in standing mode compared to using the prescribed prosthesis.

In standing, COP excursion in the anteroposterior direction was reduced in the concave standing mode compared to the convex walking mode, suggesting the relative benefit of the standing mode for stability (Fig. 9.14; one subject did not test walking mode in standing). Three of the four subjects also increased the fraction of body weight supported on the prosthetic limb using the Rock'n'Lock in standing mode compared to their prescribed prostheses, and the fourth did not change. In walking, subjects exhibited a variable relationship of metabolic energy expenditure relative to their expenditure when using their prescribed prostheses, but on average the energy expenditure was nearly identical (Fig. 9.15). These results corroborate the benefits of the mode-switching capability, with a single prosthesis having features that benefit both walking and standing behavior.

5.1.4.3 Challenges of the Rock'n'Lock design

Critical evaluation of the Rock'n'Lock prosthesis does identify some aspects that need improvement for a clinically practical version. First, the current design was described by one of the users as "thumpy" at the heel-meaning it felt too firm upon heel strike. This issue was improved in subsequent

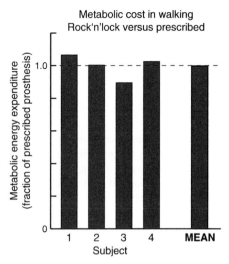

Figure 9.15 *Metabolic energy consumption during walking with the Rock'n'Lock compared to a prescribed ESR prosthesis.* Equivalent cost is surprising given that the Rock'n'Lock has no ESR capability but is simply a rigid shape.

adjustments using softer foam, but is a fundamental challenge for a system with rigid plates supporting the forefoot and hindfoot. Implementation of flexible components in these segments could improve this outcome.

Second, the Rock'n'Lock was designed for standalone use, whereas prostheses are generally used inside a shoe for daily wear. The size of the Rock'n'Lock would need to be reduced to fit inside a compatible foot shell and then inside a shoe. Once inside a foot shell and a shoe, the mechanism will not be as free to move as the shoeless prototype, so some portions of the switching mechanism would need to be redesigned to accommodate the stiffness of the added structures. The foot shell and shoe also have pre-defined shapes based on the anatomical foot, so the combined shape of the Rock'n'Lock mechanism and the compliance of the foot shell and shoe would need to be optimized together to achieve favorable characteristics for walking and standing. Nevertheless, the potential value of this mode-switching behavior is apparent from the Rock'n'Lock results.

5.2 The ankle as a spring

5.2.1 Description of behavior: increasing ankle moment with increasing angle

The second conceptual model of foot/ankle function is a lever mounted to a torsional spring. As the leg advances over the foot during walking, the angle of the ankle changes and the torque at the ankle increases

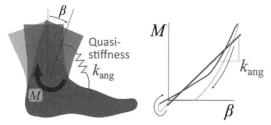

Figure 9.16 *Conceptual model of the ankle as an angular spring.* The slope of the natural ankle's moment M versus angle β curve is referred to as *quasi-stiffness*, k_{ang}; it can be defined globally as shown, or locally for different parts of the curve. Prostheses can implement similar behavior through explicit joint compliance or through deformation of hindfoot and forefoot structural keels.

proportionally; this behavior is as though the ankle were supported by a torsional spring (Fig. 9.16). This view of the ankle as a source of compliance has inspired many passive prosthesis designs, including single-axis (SA) and multi-axial (MA) ankles as well as Dynamic Elastic Response (DER) and Energy Storage and Return (ESR) foot prostheses. All these categories are designed to mimic springlike ankle behavior; the main difference is where the compliance is localized: in an articulating ankle joint (SA, MA) or in flexible keels (DER, ESR). The different structures affect the behavior of the prostheses under different loading scenarios such as vertical compression or contact at only the heel or the toe, and they affect biomechanical analysis procedures through the presence or absence of a true ankle joint. In the context of semi-active prostheses, the concept is to influence gait and balance by modifying the stiffness of the ankle spring.

5.2.2 Mathematical model: torque versus ankle angle relationship

A mathematical model of the relationship between ankle moment and ankle angle can be adapted directly from the model of a linear torsional spring. According to this model, ankle moment M_{ankle} is proportional to the angular deflection Δ_β of the joint through a torsional spring constant k_{ang}, frequently called the *ankle stiffness*: $M_{ankle} = k_{ang}\,\Delta_\beta$. The angular deflection Δ_β is defined relative to a specific neutral angle of the joint, β_0; active joints like the biological ankle and some powered prostheses can control this neutral angle as part of their control scheme. Nonlinear expressions can also be used to describe more complex moment-angle relationships.

The ankle stiffness k_{ang} can be derived from human data in multiple ways. The more accurate, but also more complex, way is to define this stiffness as one component of a multi-term equation for ankle impedance.

Impedance includes the combined effects of stiffness k_{ang} (relating ankle moment to ankle angle), damping c_{ang} (relating ankle moment to ankle angular velocity), and inertia (relating ankle moment to ankle angular acceleration). For relatively low-acceleration movements like that of the ankle during stance phase, the acceleration term is often neglected. Due to the potential for muscle action to dynamically change ankle behavior, the impedance parameters cannot be estimated from simple observational data, but must be actively probed using dynamic perturbations during movement [4,40,88-90].

A second form of ankle stiffness is termed the ankle *quasi-stiffness*: the apparent stiffness of the joint based on observed ankle moment versus angle behavior in normal movement. Quasi-stiffness can be fitted using either a differential form, $k_{ang} = dM / d\beta$, to find its instantaneous value throughout a movement [31], or using regression equations for different subparts of a movement to more succinctly describe how stiffness varies over the course of a movement [2].

In models of a springlike ankle or a prosthesis designed to mimic a spring, both inertia and damping are often neglected, leaving only stiffness to describe the mechanics of ankle movement. In this simplification, quasi-stiffness and the stiffness term of impedance are the same—simply the *ankle stiffness*. An approximate value for ankle stiffness, assuming it to be constant and linear throughout the stance phase of walking, is 0.175 dimensionless (nondimensionalized by *mgh*, mass times gravity times height of an individual; estimated from mean data in [2] table 1, plantarflexion phase or from [4] early stance; typical dimensional values 200-300 Nm/rad).

5.2.3 Variations: effects of different stiffnesses

Stiffness in the natural ankle (either quasi-stiffness or impedance) varies in complex ways throughout the stance phase and with the speed of walking [91]. This real-time adjustment of stiffness during loaded phases of gait requires the addition of substantial mechanical power from the muscles; to fully mimic this behavior with a prosthesis would require a high-power actuator (this class of actuator is called a *variable-stiffness actuator*, e.g. [92-96]). However, the cyclic loading and unloading of the leg in human movement offers an opportunity for once-per-stride modulation of the prosthesis' stiffness, in which stiffness is adjusted in response to the movement and environment during unloaded swing phases. This type of semi-active design may benefit the user through different mechanisms from those found in the natural ankle.

The potential impacts of changing prosthesis stiffness have been investigated in theoretical and experimental studies. One key theoretical study investigated the hypothesis that ankle stiffness and foot length interact to determine the timing and magnitude of ankle push-off and its effects on energy expenditure [37]. In this study a parametric dynamic walking model suggested that for any foot length there is an optimal stiffness, and for any stiffness an optimal foot length, to minimize the mechanical energy needed to maintain walking. The energy savings was based on the timing and magnitude of energy return at the end of each stance phase: a stiffer ankle caused earlier timing, but also lower magnitude of energy return. The trade-off resulted in an optimal stiffness for a given foot length. An experimental study of the effects of prosthesis hindfoot and forefoot stiffness at different speeds found similar effects of varying forefoot stiffness: lower stiffness led to increased magnitude but later timing of energy return [28] (Fig. 9.17). Different stiffnesses of the hindfoot and forefoot also affected the posture, moment and power output of the ankle and knee, but had minimal effect on the hip, suggesting that the knee resolves most of the effects of a below-knee prosthesis [28]. Related experiments with prostheses of different overall stiffness showed similar effects on prosthesis energy return and knee mechanics [47]. Analysis of these results through musculoskeletal modeling suggested specific changes in muscle function [97] that could allow optimization of prosthesis stiffness to compensate for specific injuries. Both experimental studies suggested that a stiffer prosthesis (likely the forefoot) leads to increased vertical ground forces on the intact side [28,47].

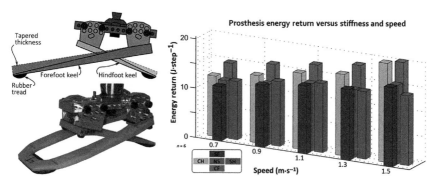

Figure 9.17 *A study of independent variations of hindfoot and forefoot stiffness in an experimental prosthesis showed that reducing component stiffness—especially forefoot stiffness—increases the mechanical energy returned to the user. NS: nominal stiffness; CH: compliant hindfoot; SH: stiff hindfoot; CF: compliant forefoot; SF: stiff forefoot. Figure adapted from Ref. [28].*

Those studies showed effects of changing stiffness on gait biomechanics, but that simply argues for careful tuning at the time of prescription. Is there any reason to desire real-time control through a semi-active mechanism? Several lines of evidence suggest there is. First, it is common knowledge to prosthetists and persons with amputation that walking on slopes is uncomfortable and requires strange biomechanical adaptations when using ordinary prostheses. A semi-active device could adapt to such conditions by rebalancing the relative stiffness of the forefoot and hindfoot sections. This idea is supported by findings of the ill effects of prosthesis misalignment (similar to sloped ground) [98-100]. Second, a study of gait in persons with amputation using typical ESR prostheses compared to unimpaired persons revealed a significant shortcoming of the prostheses: the natural leg substantially modulates its push-off power (largely provided by the ankle) as walking speed changes, but the amount of energy returned by prostheses does not change with speed [101]. In combination with the findings from changes in prosthesis stiffness, an opportunity arises to use semi-active changes in prosthesis stiffness to overcome this lack of modulation in energy storage and return. Finally, the difference between walking and standing can be addressed through changes in stiffness, instead of changes in shape [87,102].

Two recent semi-active prostheses have been developed to implement control of prosthesis stiffness based on variations of a variable-length cantilever beam [23,58]. This section presents the key design considerations for one of them, a design that retains the typical flexible-keel architecture of most prostheses but adds real-time variation of prosthesis stiffness, all while maintaining very low prosthesis mass, build height and power consumption [58].

5.2.4 Implementation: the variable-stiffness foot (VSF)

The Variable-Stiffness Foot (VSF) is designed to vary the stiffness of the forefoot section of the prosthesis across a roughly fourfold range. The forefoot's structural keel is chosen as the variable-stiffness member because it is the primary energy storage and return (ESR) element in many high-performance passive prostheses. Therefore the overall mechanics of the VSF prosthesis will mimic those of existing ESR prostheses. But, the ability to control stiffness of the forefoot allows the VSF to control the energy storage and return capacity of the prosthesis, which standard prostheses cannot do. Stiffness control also influences the sagittal plane moment on the prosthetic socket, which is an important quantity affecting discomfort from stresses on the residual limb.

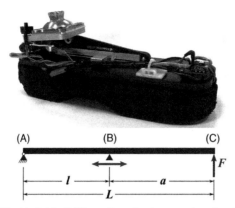

Figure 9.18 (Top) *The Variable-Stiffness Prosthesis, a semi-active device that controls forefoot stiffness on a stride-by-stride basis.* **(Bottom)** *Overhung Euler beam model of the Variable-Stiffness Prosthesis.* When external ground reaction force F is applied at the forefoot, the effective stiffness is controlled by the overall length L and overhung length a that protrudes past the movable fulcrum (B), and the cross section and modulus of the beam. *Adapted from Ref. [58].*

The variable-stiffness feature is implemented by designing the forefoot keel as a flexible Euler-Bernoulli beam with a pin connection at the posterior end and the forefoot overhanging a moveable support fulcrum [58]. Fig. 9.18 shows the VSF prosthesis with a schematic beam model to illustrate this function. Under a force F applied at the toe (point C), the vertical stiffness k_v at the contact point is determined by the keel length L, the modulus E of the keel material, the moment of inertia in bending, I, and the overhang length a [58]:

$$k_v = 3EI / (La^2). \qquad (9.1)$$

This relationship shows that the vertical stiffness increases as the overhang length decreases. In the VSF design, keel parameters and the range of overhang lengths a were chosen with the goal of achieving a factor of four in stiffness variation. A similar relationship describes the variation in stiffness at any forefoot contact point, with appropriate values of L and a.

Equation (9.1) describes the vertical stiffness of the forefoot, which is a surrogate for angular ankle stiffness k_{ang} when only the forefoot is in contact with the ground:

$$k_{ang} = d^2 k_{v\_fore}, \qquad (9.2)$$

where d is the forward distance from the ankle to the forefoot contact point. During foot-flat, angular stiffness is affected by both hindfoot and forefoot contact. The relationship depends on the distance l between the hindfoot and forefoot contact points as well as the vertical stiffness at each point, k_{v_hind} and k_{v_fore} [42]:

$$k_{ang} = l^2 k_{v\_hind} k_{v\_fore} / \left(k_{v\_hind} + k_{v\_fore} \right). \qquad (9.3)$$

This relationship is complex, but it is monotonically increasing with forefoot stiffness. Therefore control of forefoot stiffness provides a means of controlling the angular stiffness of the whole prosthesis.

5.2.4.1 Semi-active design features of the VSF

The VSF mechanism exploits two critical semi-active design features: near-orthogonal actuation and a friction-based non-backdrivable mechanism. In the bending beam model of Fig. 9.18, displacement of the keel is primarily vertical, but the motion of the fulcrum is horizontal. Because of this near-orthogonal arrangement, the support force from the moveable fulcrum is also nearly vertical and very little horizontal force is required to hold it in place. Therefore the fulcrum can be mounted on a slider and left free to move horizontally, while being supported rigidly against vertical displacement. When no external force is applied, the long axis of the keel aligns with the motion of the fulcrum, allowing it to move freely. When external force is applied, the keel pushes the fulcrum vertically against its supports, leading to a large friction force in the slider to hold it in place.

At large forces the deflection of the keel is more substantial and the near-orthogonality assumption breaks down due to the slope of the bending beam at the point where the fulcrum supports it. This angle requires the fulcrum to provide a component of the support force along the horizontal slider axis, for which slider friction may not be sufficient (see Fig. 9.6). To prevent backdriving of the fulcrum in this case, the fulcrum is provided with urethane friction pads at its contact surface with the keel (Fig. 9.7). When there is no external force, the pads are unloaded and provide minimal resistance to the free motion of the fulcrum, but when loaded their high friction coefficient provides additional holding capacity to prevent backdriving.

The motion of the fulcrum is actuated by a miniature gear motor with a timing-belt drive, driven by an embedded microcontroller. The microcontroller includes an inertial measurement unit (IMU) that measures the motion of the prosthesis and reconstructs the position and velocity of the foot [58,61]. Periods of unloading, when the fulcrum is free to move, are identified by high angular velocity (see Fig. 9.3).

5.2.4.2 Performance and benefits of the VSF

The VSF achieves a factor of 3.2 in forefoot stiffness modulation as measured using a material test machine (k_{v_fore} = 10 to 32 kN/m). The heel has stiffness k_{v_hind} = 66 kN / m , attributable only to compression of the EVA foam sole material. The ankle-to-forefoot distance d is roughly 0.15 m and the hindfoot-to-forefoot contact distance l is roughly 0.16 m, leading to estimated angular stiffness during forefoot-only support ranging from 225 to 720 Nm/rad (3.9 to 12.6 Nm/deg) and during foot-flat from 222 to 552 Nm/rad (3.9 to 9.6 Nm/deg). The VSF fulcrum is able to move through the full range of stiffness settings in roughly 2.5 strides, likely faster than it will need to move in practical use.

Control of forefoot stiffness successfully leads to modulation of energy storage and return, as measured through initial human subjects testing. Energy storage and return are 60-80% greater in the most compliant setting compared to the stiffest setting (Fig. 9.19). Therefore the VSF mechanism could potentially make adjustments to achieve more humanlike changes in walking speed by lowering the forefoot stiffness as speed increases. Similarly, the VSF could increase to its stiffest settings at low or zero walking speed to approximate the high stiffness values used in standing.

Ongoing testing seeks to establish the effects of the VSF on users when stiffness is being actively controlled. The biomechanical benefits to the proximal knee and contralateral leg are important outcomes that will be evaluated across speeds and slopes. Additionally, the preferred pattern of stiffness modulation with speed and slope is important, and will inform practical application of the technology.

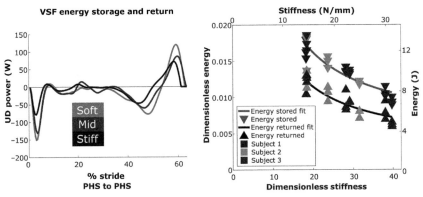

Figure 9.19 (Left) *Example prosthetic side uniform deformable segment power delivered to the shank by the VSF prosthesis for walking speed at 1.2 m/s.* (Right) *The quantity of energy stored and returned increases with lower stiffness settings.* Adapted from Ref. [58].

5.2.4.3 Challenges of the VSF design

As with the Rock'n'Lock prosthesis, one challenge is to fit the VSF within a foot shell and a shoe. The presence of moving components in distal areas of the foot presents a potential failure point if the foot shell becomes damaged or collects debris. Another challenge is to define clearly the circumstances under which the prosthesis should adjust its stiffness, and in what way it should do so. Programming the devices to respond appropriately and predictably to human behavior is an area of ongoing research for virtually all active prostheses, including semi-active prostheses.

5.3 The ankle as an angle adapter

5.3.1 Description of behavior: ankle adaptation to slopes, stairs, speeds, and turns

The third conceptual model of ankle function is a controller for the neutral angle of a passive foot. During different types of movement or in locomotion on different terrains, the ankle articulates to different angles relative to the leg, often with dramatically different joint torques than would be predicted by the spring or rollover shape models. Such gross angle changes are possible because the muscles that control the ankle can control force and displacement independently. This capability allows the ankle joint to operate with spring-like or shape-like behavior relative to a freely chosen neutral angle.

This change in the neutral ankle angle is the basis for several commercially available passive and semi-active prostheses. Some ankle modules use passive hydraulic damping to allow the ankle angle to change in response to sustained loading, achieving a stress-relaxation behavior that reduces moments on the residual limb [15,25,103-105]. These systems allow the ankle angle to change and mount a deformable foot prosthesis beneath it to achieve springlike foot mechanics. Similar mechanisms with semi-active control of the damping are the most popular form of microprocessor-controlled prosthesis [13,16,18,19]. Finally, one robotic prosthesis semi-actively controls the angle of the ankle, again with a deformable foot underneath [20].

These systems are effective in the sagittal plane, but the ankle moves in the frontal and transverse planes as well. The frontal plane is particularly important, for two reasons: locomotion is inherently less stable in the lateral direction (frontal plane toppling) than in the sagittal plane [106]; and terrain encountered in real-world locomotion is uneven in both the frontal and sagittal planes, making adaptation in both directions desirable. Control of the ankle in the frontal plane is an important component of balance control [107,108], and under-foot disturbances in the frontal plane often require

aggressive stepping strategies for recovery [109,110]. Despite the importance of frontal plane motion, all currently available prostheses implement
frontal plane angular motion only by spring-like compliance from a fixed
neutral angle. Even in research, only one prosthesis implements a clutching mechanism to allow frontal displacement at ground contact [111], and
a few others use fully powered control of two-plane motion [112-114]. In
contrast to these approaches, a semi-active design built on the concept of
controlling neutral angle would actively change frontal and/or sagittal ankle
angle during unloaded swing phases, and allow the stiffness and shape of the
prosthetic foot underneath to manage the forces.

5.3.2 Mathematical model: controlling the neutral angle of a spring or shape

The ankle's neutral angle is a kinematic quantity, so it does not directly affect ankle loads. Instead, neutral angle interacts with the shape and stiffness
to determine ankle mechanical behavior. Therefore, a mathematical model
of neutral angle control must include these other model components. Fig.
9.20 shows the effects of a kinematic change at the ankle assuming a constant rollover shape or a constant ankle stiffness for the foot. First, if the
foot has a constant circular arc shape, then changing the ankle angle moves
the arc shape with the foot. The mathematics of this motion simply the reflect the change in coordinates of the circle center due to a rotation about
the ankle. Fig. 9.20 (left) shows the typical rollover shape in level walking
(light shading), and an altered rollover shape after the ankle angle changes
by an angle β_0. If the vector from the ankle center to the rollover shape's
arc center has length ρ and is typically at an angle γ from vertical, then
after the change its angle is $\gamma + \beta_0$. Thus, the anterior (x) and vertical
(Y) coordinates are simply defined by the angle: $x = \delta \sin(\gamma + \beta_0)$ and
$y = \delta \cos(\gamma + \beta_0)$. If this new arc shape is characterized again with respect
to a leg-fixed reference frame centered at the ankle, the effect viewed from
outside is simply a translation of the observed rollover shape in the vertical
and anteroposterior directions by distances $\Delta x = \delta \sin(\gamma + \beta_0) - \delta \sin(\gamma)$
and $\Delta y = \delta \cos(\gamma + \beta_0) - \delta \cos(\gamma)$. Thus, the rollover shape model predicts that the ankle angle can control the vertical and forward offsets of the
observed rollover shape. Since rollover shape is estimated from the center of
pressure, this prediction implies that the COP moves farther forward with
a more plantarflexed ankle.

   If instead the foot-ankle mechanics are modeled as a spring, then the
change in ankle angle simply affects the effective spring displacement

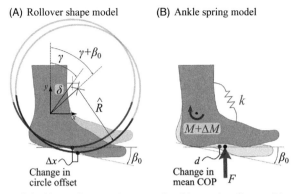

(A) Rollover shape model   **(B)** Ankle spring model

Change in
circle offset

Change in
mean COP

Figure 9.20 *Model of how changes in neutral ankle angle affect ankle mechanics assuming a rollover shape or ankle stiffness model distal to the articulated joint.*

at any absolute ankle angle—and therefore the resulting ankle moment. Fig. 9.20 (right) shows a case in which the neutral angle is plantarflexed by an angle β_0, and from this neutral position the forefoot has been rotated upward again by a vertical ground reaction force F. The moment necessary to rotate the ankle through this angle is proportional to the ankle stiffness: $\Delta M = k\,\beta_0$. This case is compared to another that has the same final configuration and ground force but a less plantarflexed neutral angle: the ankle moment increases from M to $M + \Delta M$ in order to deflect the spring, and the location of the center of pressure (COP) through which the force acts is changed by a distance $d = \Delta M / F = k\,\beta_0 / F$. Thus, the ankle stiffness model predicts that the ankle angle can control the forward offset of the center of pressure. This prediction is consistent with the prediction of a COP shift based on rollover shape. Therefore both models illustrate the ability of ankle angle shifts to control the location of ground force interaction: plantarflexion to push the COP toward the toes, and dorsiflexion to push the COP toward the heels.

The stiffness model has a correlate in the frontal plane. Because the human ankle and any passive prosthesis has an effective stiffness in inversion/eversion [39,40], a similar derivation leads to similar conclusions. Therefore changes in the frontal plane neutral angle are predicted to lead to movements of the COP in the mediolateral direction.

5.3.3 Variations: effects of angle changes

It is challenging to measure the effects of ankle angle in natural movement due to the inability to specify it as an independent variable in walking experiments. However, certain clues are available based on different movement conditions that have been observed. For example, experiments of slope walking in intact individuals show that the ankle angle during stance changes in the same direction as the slope of the ground and the moment does not scale directly as it would with a constant neutral angle [115-118]. Similarly, cross-slope walking leads to changes in frontal-plane hind-foot and forefoot angles to match the frontal slope [119]. Another approach is to observe changes in center of pressure across activities. For example, one study found that increases in walking speed lead to concentrations of pressure more on the hindfoot and less on the forefoot [120]. Recent work by the author found this same shift in COP, as well as a jump forward in COP when switching to a running gait [121]. These results, together with the mathematical models above, suggest that the angle's neutral angle is indeed being controlled.

The effects of deliberate angle changes in a prosthetic leg have been studied, but typically using adjustments at the socket rather than the ankle. Several studies have shown the effects of alignment changes on socket moment, which relates to ankle moment, including both sagittal and frontal plane effects [98-100,122]. A related study showed that translation of the foot prostheses also affects moments in the leg [123], including cross-axis coupling between sagittal and frontal plane moments. Finally, a classic study on aligning prostheses based on rollover shape demonstrated that rotating and translating a prosthesis can move the rollover shape relative to the leg [83], confirming the mathematical model relating ankle angle and rollover shape.

5.3.4 Implementation: The Two-Axis 'Daptable Ankle (TADA)

The Two-Axis 'Daptable Ankle (TADA) [124] is designed to control the neutral angle of the prosthesis in both the frontal and sagittal planes, with a total inclination angle up to ± 10 degrees in any direction (Fig. 9.21). The TADA is built as an independent module to allow installation with any foot prosthesis, allowing the mechanics of the prosthesis itself to remain unchanged. The goal is to add controlled movement of the ankle to allow adaptation to different slopes of terrain including side-slopes. Angle control in the frontal plane also has the potential to augment turning, walking balance, speed changes, and the transition between walking and running.

The two-axis angle control uses a novel mechanism consisting of a universal joint (two-axis rotational joint) nested within a system of rotating cam wedges (Fig. 9.22). The universal joint allows plantarflexion/dorsiflexion and inversion/eversion movements (PF/DF and IV/EV) while supporting axial moments to prevent internal/external rotation of the foot, and also

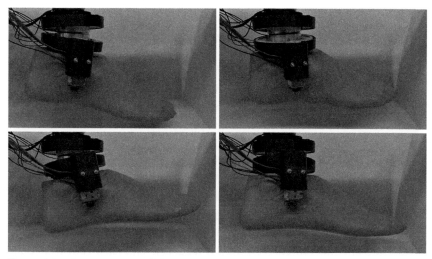

Figure 9.21 *Extreme poses of the Two-Axis 'Daptable Ankle (TADA) in a benchtop test: 10 degrees plantarflexion* (**Top Left**) *and Dorsiflexion* (**Bottom Left**)*; 10 degrees inversion* (**Top Right**) *and eversion* (**Bottom Right**). Angles are observed most clearly at the black housing. ± 10 degrees range of motion can be achieved about any combination of these PF/DF and IV/EV axes. Future design revisions will further miniaturize and integrate the motors into the compact ankle module.

Two-Axis Daptive Ankle (TADA)

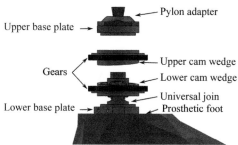

Figure 9.22 *Structure of the TADA mechanism.* A two-axis Universal joint prevents the prosthesis from rotating about the leg axis, while allowing PF/DF and IV/EV movement. Two "cam wedges" with inclined mating faces rotate about this vertical axis, pushing on each other to manipulate the foot in both motions. *Adapted from Ref. [124].*

supporting reaction forces in all directions (axial tension, anteroposterior shear and mediolateral shear). The cam wedge mechanism consists of two short cylindrical components free to rotate about a cylindrical axis aligned with the longitudinal axis of the universal joint. The cylindrical components have their mating faces inclined at angle γ (here $\gamma = 5$ degrees) along a plane through the center of the universal joint, giving each the appearance of a wedge. These wedges have gear teeth mounted around their perimeter, which are driven by small motors to cause rotation about the central axis. When one wedge turns, its inclined face always makes an angle γ with the vertical axis, but the direction of that inclination rotates with the wedge. The second wedge has a similar effect, which can be used to cancel the upper rotation (a neutral ankle angle), add to it in the same direction (full inclination of 2γ), or add to it in a different direction (total inclination angle between 0 and 2γ, in a direction between the two wedge faces). This behavior is effectively that of a cam system, so these wedge components are called "cam wedges." The cam wedges are actuated using small DC gear motors with pinion gears, all contained within upper and lower housings attached to the respective base plates. The motors have magnetic quadrature encoders to enable position feedback, and each cam wedge has a small permanent magnet installed to register absolute position on a Hall effect sensor mounted on the base.

The kinematics of the system (Fig. 9.23) are based on the intuition that the direction of net downward inclination, at angle α from forward, is halfway between the directions of the two wedge faces, and the total inclination angle θ depends on the magnitude of the angle φ between the two wedges' orientations, φ_1 and φ_2. Mathematically, the relationships are:

$$\varphi = \varphi_2 - \varphi_1 \tag{9.4}$$

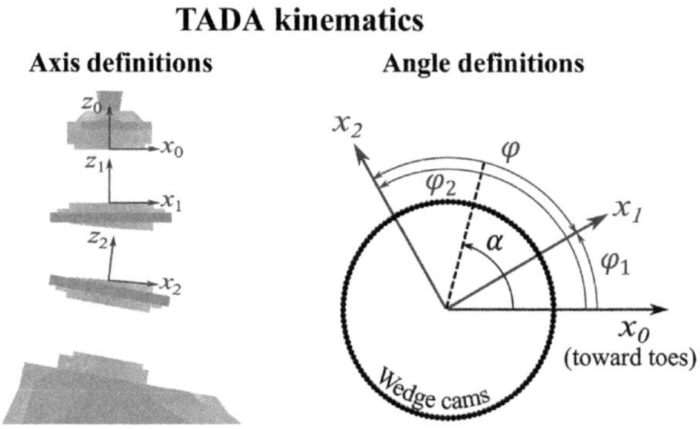

TADA kinematics

Axis definitions   Angle definitions

Figure 9.23 *Kinematics of the TADA ankle mechanism.* (*Left*) Axis definitions: The x axis of each wedge cam points from its center toward its thickest point. The x_0 axis is fixed to the shank and points toward the toes. (*Right*) Angles φ_1 and φ_2 are measured from the x_0 axis to the x_1 and x_2 axes, respectively. Angle φ is the difference between them and the overall "downward direction" α is their mean. Control of the prosthesis is determined by a desired downward direction α and inclination θ that determines φ.

$$\alpha = \left(\varphi_2 + \varphi_1\right)/2 \tag{9.5}$$

$$\theta = \cos^{-1}\left(\cos^2\left(\gamma\right) - \cos\left(\varphi\right)\sin^2\left(\gamma\right)\right) \tag{9.6}$$

These relationships can be inverted to solve for φ_1 and φ_2 when θ and α are specified; then φ_1 and φ_2 become the control targets for the two cam wedges' axial rotation angles:

$$\varphi = \cos^{-1}\left(\frac{\cos^2\left(\gamma\right) - \cos\left(\theta\right)}{\sin^2\left(\gamma\right)}\right) \tag{9.7}$$

$$\varphi_1 = \alpha - \varphi/2 \tag{9.8}$$

$$\varphi_2 = \alpha + \varphi/2 \tag{9.9}$$

The order of the cam wedges (1 or 2) is not important, but it is important that the angles φ_1 and φ_2 are defined by the thickest point of the wedge, and φ_1, φ_2, and α are all measured relative to the anterior anatomical direction (forward foot axis). A more traditional derivation based on serial-chain robotic manipulator kinematics can also be used to directly relate these rotations to explicit plantarflexion and inversion angles [124]; however, the resulting expressions are quite complex.

5.3.4.1 Semi-active design features of the TADA

The TADA exploits the semi-active principles of nonbackdriveability and swing-phase actuation to minimize weight and power. The inclined faces of the cam wedges sustain a large compressive force to support the ankle moment resulting from ground contact forces acting elsewhere on the foot—in the worst case, at the toe. The compressive force has a component normal to the surface of the wedge, and this component creates a back-driving twist moment that tends to rotate the cams. However, the mechanism is built so that the compressive stresses acting at the interface between the two cams, together with the stresses at the interface between each cam and the base plate it touches, create enough friction force to resist the backdriving moment. Using friction on both faces of the cam wedge doubles the resistive friction torque relative to the simple inclined plane. Therefore the estimated minimum friction coefficient $\mu = 0.1$ is sufficient to support wedge angles up to $\gamma = 11.4$ degrees—a safety factor greater than two relative to the designed wedge angle of five degrees. Thus, the cam wedges do not need to be held in position actively—they implement set-it-and-forget-it control. This mechanism is simpler to build than a conventional platform that might use a universal joint and two linear actuators [112], [114], and potentially lighter due to elimination of heavy power screws and the associated bearings.

The compressive contact forces at the interface are orders of magnitude higher during stance phases than during swing phases. Therefore the friction force that holds the cams in position is mostly eliminated when the foot is off the ground. Because resistance to cam rotation is minimal in this case, the TADA can reposition the cams using small, lightweight, low-power motors. These motors operate under position feedback from their own quadrature encoders, which relates directly to the angles of the cam wedges. However, the TADA does not use traditional linear (e.g. proportional) position control, because of the remaining friction in the system. Proportional control in a system with friction cannot guarantee convergence and can waste electrical energy if the system arrives at a configuration where the resulting command is insufficient to overcome friction. A proportional controller would continue driving current through the motor, but not enough to make it move to reduce the error. To prevent this problem, the TADA uses a variation of bang-bang control, in which the motor is powered at ±100% whenever the error exceeds a threshold value, and is shut off when the error is within this threshold (the so-called dead-band). In this manner the motor is never fighting friction with an insufficient torque command. Because the motor command acts through an inertial system, the TADA controller shifts the dead-band toward the direction of approach, to shut off the motor slightly sooner and achieve better position accuracy.

5.3.4.2 Performance of the TADA

The TADA achieves errors in cam wedge orientation less than 3.4 degrees in 99% of cases, which result in ankle angle error magnitudes less than 0.5 degrees on average in both PF/DF and IV/EV (Fig. 9.24). All movements of the mechanism can be achieved in less than 0.5 seconds, with roughly 95% in less than 0.3 seconds, ensuring that a single swing phase is adequate to move from any pose to any other pose. Use of an ankle module rather than a custom-built prosthesis means that fitting into shoes is more readily achievable compared to the Rock'n'Lock and VSF prostheses, though further work is still needed to improve this capability.

Human subjects testing with the TADA has not yet been undertaken at the time of publication, so its effects on walking and other activities are not

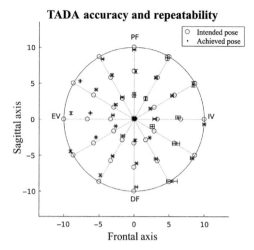

Figure 9.24 *Positioning error and variability of the TADA over 240 randomized configuration changes.* Mean error is less than 0.5 degrees in either direction.

yet known. Tests planned include: determining the effects of static alignment changes at the ankle joint; testing the ability to reduce socket moments on sloped terrain through environment-aware adaptation of ankle angle; determining the effects of sudden changes in alignment on gait mechanics; and using the TADA for perturbation training to improve balance.

5.3.4.3 Challenges of the TADA design

Backlash in the mechanism is the most serious limitation of the TADA, with magnitude up to 1.6 deg PF/DF and 0.98 deg IV/EV. The direct cause of this backlash is tolerance stack-up in the mechanism, but it is fundamentally attributable to the fact that the mechanism is formally over-constrained. In order to function correctly, the geometric center of the interfacing surface between the two wedge cams must align perfectly with the joint center of the two-axis universal joint. Imperfect alignment is inevitable, and must be dealt with by allowing some backlash in the movement. However, improved tolerancing and dimensional stability of the articulating parts could reduce the problem in future revisions.

Another challenge is coordinating the motion of the two cams. Because the two cam wedges can independently drive the ankle in all directions, poorly planned or poorly synchronized movement can result in unfavorable intermediate poses during the transition from one pose to another. For example, in transitioning from inversion to eversion, if both cam wedges drive through angles near zero (with the thickest edge anterior) the foot will be near 10 degrees plantarflexion, presenting a trip hazard during swing phase. Or, if one cam wedge moves very little but the other needs to move much further, this combination can drive the foot in a circumduction motion, again presenting a trip hazard and also looking strange to an observer. These problems have been improved by using synchronized interpolated poses aimed at achieving well-controlled movements of the foot and through heuristic rules to eliminate specific undesirable movements, but more work is still required to optimize the movements of the TADA for safety, smoothness and quickness.

Finally, the geometric placement of the ankle is a challenge affecting the hypothesized effects and benefits of any two-axis ankle. In the frontal plane, inversion of the ankle is intended to move the center of pressure toward the lateral edge of the foot prosthesis. But, this COP motion only happens if the ankle axis is relatively low. If the ankle is high, then the angular inversion of the foot sole is counteracted by overall movement of the prosthesis toward the mid-line of the body. Therefore care must be taken to place the ankle as low as possible to allow the frontal plane motion to have the intended effect. An alternative possibility may be to use the TADA at the socket instead of the ankle, which is less biomimetic but perhaps equally useful for enhancing foot placement.

6 Challenges of semi-active prostheses

The examples of the Rock'n'Lock, the VSF and the TADA illustrate several challenges that are particularly prevalent in semi-active prostheses. First, the goal of saving weight and reducing power consumptions leads to designs with minimal actuation using very small motors and other miniature actuators. Unfortunately, these miniature actuators are more susceptible to environmental and structural disturbances that can prevent them from moving at all, compared to higher-power actuators that can overcome the same disturbances. Therefore, care should be taken to design mechanisms that circumvent this risk, and to avoid under-sizing the actuators. Often the choice of actuator sizing is a process of trial-and-error, since the problematic phenomena like friction, debris build-up and tolerance stack-up are difficult to model predictively. Systems that enclose the mechanism in a protective housing can reduce the risk of this kind of problem.

A second challenge is that the dependence of semi-active prostheses on specific external conditions of loading and movement renders them vulnerable to abnormal or unexpected behaviors by the user. For example, the Rock'n'Lock prosthesis requires unloading in order to move into standing mode, but a user may not always lift her foot when she stops walking. Similarly, the VSF and TADA prostheses both move only during swing phases, and can therefore fail to adapt if the user does not lift and move her foot. And, the CESR foot will fail to release its toe latch if the forefoot force is not sufficient. A related challenge is that these different movement conditions must be sensed and classified and a control decision must be made, all of which can be problematic for unusual movements. For example, how should the VSF detect and respond to swinging on a swing, riding on a bouncy bus, or ice skating? In these cases motion may exceed the programmed threshold, but in swinging the VSF should not take any action in response and on the bus any attempt to move would fail because the foot is not unloaded. Detecting and handling these and other exceptions can be a complex exercise in controller programming.

Finally, the nature of any active or semi-active system is that its properties are time varying, and this variation can create problems if users are unable to predict it. This challenge is related to the larger issue of human-robot interaction, which the user's intent must be communicate to an external active system. Current solutions are limited; strategies used include defining specific movements through which a user explicitly cues the system to switch modes (e.g. the Össur Proprio Foot's stairs mode), or activity classification techniques based on motion and load on the prosthesis [125-127]. Use of predefined movements is generally disliked by users because it is not intuitive. Research to improve activity classification includes adding signals such as electromyographic signals or environmental sensing [8,128-130] or implementing direct myoelectric control [131,132]. Creating reliable control that makes active devices predictable and intuitive for the user is an ongoing challenge.

7 Conclusion

The era of biomechanically responsive prosthetics is well underway. Semi-active prostheses are likely to remain key players in the prosthetics market, as users seek greater adaptability and improved function without the weight, height, complexity and price of fully-powered devices. The key enablers of semi-active devices, including clutching, non-backdriveability, use of springs, and exploitation of the gait cycle, can help designers effectively translate concepts into practicable designs of future biomimetic and non-biomimetic prostheses.

References

[1] A.R. De Asha, R. Munjal, J. Kulkarni, J.G. Buckley, Walking speed related joint kinetic alterations in trans-tibial amputees: impact of hydraulic ankle' damping, J. NeuroEng. Rehab. 10 (2013) 107 doi: 10.1186/1743-0003-10-107.

[2] K. Shamaei, G.S. Sawicki, A.M. Dollar, Estimation of Quasi-stiffness and propulsive work of the human ankle in the stance phase of walking, PLoS ONE 8 (3) (Mar 2013) e59935 doi: 10.1371/journal.pone.0059935.

[3] L. Jin, M.E. Hahn, Modulation of lower extremity joint stiffness, work and power at different walking and running speeds, Human Mov. Sci. 58 (2018) 1–9 doi: 10.1016/j.humov.2018.01.004.

[4] E.J. Rouse, L.J. Hargrove, E.J. Perreault, T.A. Kuiken, Estimation of human ankle impedance during the stance phase of walking, IEEE Trans. Neur. Sys. Rehab. Eng. 22 (4) (2014) 870–878 doi: 10.1109/TNSRE. 2014.2307256.

[5] M. Cempini, L.J. Hargrove, and T. Lenzi. Design, development, and bench-top testing of a powered polycentric ankle prosthesis, in: 2017 IEEE/RSJ International Conference on Intelligent Robots and Systems (IROS), 2017, pp. 1064-1069, doi: 10.1109/IROS.2017.8202276.

[6] T. Lenzi, M. Cempini, J. Newkirk, L.J. Hargrove, and T. A. Kuiken. A lightweight robotic ankle prosthesis with non-backdrivable cam-based transmission, in: 2017 International Conference on Rehabilitation Robotics (ICORR), 2017, pp. 1142-1147, doi: 10.1109/ICORR. 2017.8009403.

[7] A. Parri, et al. Whole body awareness for controlling a robotic transfemoral prosthesis, Front. Neurorobot. 11 (2017) 25 doi: 10.3389/fnbot.2017.00025.

[8] A.J. Young, T.A. Kuiken, L.J. Hargrove, Analysis of using EMG and mechanical sensors to enhance intent recognition in powered lower limb prostheses, J. Neural Eng. 11 (5) (2014) 056021 doi: 10.1088/1741-2560/11/5/056021.

[9] C-Leg 4 above knee prosthetic leg. [Online]. Available from: https://www.ottobockus.com/prosthetics/lower-limb-prosthetics/solution-overview/c-leg-above-knee-system/. [Accessed: 22-Mar-2019].

[10] Genium leg prosthesis | Ottobock US. [Online]. Available from: https://www.ottobockus.com/prosthetics/lower-limb-prosthetics/solution-overview/genium-above-knee-system/. [Accessed: 22-Mar-2019].

[11] X3 waterproof prosthetic leg | Ottobock US. [Online]. Available from: https://www.ottobockus.com/prosthetics/lower-limb-prosthetics/solution-overview/x3-prosthetic-leg/. [Accessed: 22-Mar-2019].

[12] Össur Rheo Knee, Össur Rheo Knee. [Online]. Available from: http://www.ossur.com/prosthetic-solutions/products/dynamic-solutions/rheo-knee-3. [Accessed: 22-Aug-2016].

[13] Élan. [Online]. Available from: http://www.endolite.com/products/elan. [Accessed: 19-Jun-2018].

[14] OdysseyK2. [Online]. Available from: https://www.college-park.com/odysseyk2. [Accessed: 19-Jun-2018].

[15] OdysseyK3. [Online]. Available from: https://www.college-park.com/odysseyk3. [Accessed: 19-Jun-2018].

[16] Raize. [Online]. Available from: http://fillauer.com/Lower-Extremity-Prosthetics/feet/raize.html. [Accessed: 19-Jun-2018].

[17] Meridium. [Online]. Available from: https://www.ottobockus.com/prosthetics/lower-limb-prosthetics/solution-overview/meridium/. [Accessed: 12-May-2018].

[18] Triton smart ankle. [Online]. Available from: https://www.ottobockus.com/prosthetics/lower-limb-prosthetics/solution-overview/triton-smart-ankle/. [Accessed: 19-Jun-2018].

[19] Kinnex, Freedom Innovations, 13-Oct-2016. [Online]. Available: http://www.free-dom-innovations.com/kinnex/. [Accessed: 19-Jun-2018].

[20] "Össur Proprio Foot," Össur Proprio Foot. [Online]. Available: http://www.ossur.com/prosthetic-solutions/bionic-technology/proprio-foot. [Accessed: 13-Sep-2014].

[21] P. G. Adamczyk. The influence of center of mass velocity redirection on mechanical and metabolic performance during walking, Ph.D., University of Michigan, United States -- Michigan, 2008.

[22] A. H. Hansen and D. S. Childress. Bi-modal ankle-foot device. US8764850 B2, 01-Jul-2014.

[23] M.K. Shepherd, E.J. Rouse, The VSPA foot: a Quasi-passive ankle-foot prosthesis with continuously variable stiffness, IEEE Trans. Neural Sys. Rehab. Eng. 25 (12) (Dec 2017) 2375–2386 doi: 10.1109/TNSRE.2017.2750113.

[24] J. Zhu, Q. Wang, and L. Wang. PANTOE 1: Biomechanical design of powered ankle-foot prosthesis with compliant joints and segmented foot, in: Advanced Intelligent Mechatronics (AIM), 2010 IEEE/ASME International Conference on, 2010, pp. 31-36.

[25] "Echelon," Endolite USA - Lower Limb Prosthetics. [Online]. Available: http://www.endolite.com/products/echelon. [Accessed: 08-Jun-2018].

[26] A.R. De Asha, R. Munjal, J. Kulkarni, J.G. Buckley, Impact on the biomechanics of overground gait of using an 'Echelon' hydraulic ankle-foot device in unilateral trans-tibial and trans-femoral amputees, Clin. Biomech. 29 (7) (2014) 728–734 doi: 10.1016/j.clinbiomech.2014.06.009.

[27] C.-Y. Ko, et al. Biomechanical features of level walking by transtibial amputees wearing prosthetic feet with and without adaptive ankles, J. Mech. Sci. Technol. 30 (6) (2016) 2907–2914 doi: 10.1007/s12206-016-0550-6.

[28] P.G. Adamczyk, M. Roland, M.E. Hahn, Sensitivity of biomechanical outcomes to independent variations of hindfoot and forefoot stiffness in foot prostheses, Human Move. Sci. 54 (2017) 154–171 doi: 10.1016/j.humov.2017.04.005.

[29] E. J. Rouse, L.M. Mooney, E.C. Martinez-Villalpando, and H. M. Herr. Clutchable series-elastic actuator: design of a robotic knee prosthesis for minimum energy consumption, in: 2013 IEEE 13th International Conference on Rehabilitation Robotics (ICORR), 2013, pp. 1-6, doi: 10.1109/ICORR.2013.6650383.

[30] S.H. Collins, A.D. Kuo, Recycling energy to restore impaired ankle function during human walking, PLoS One 5 (2010) e9307.

[31] E. Singer, G. Ishai, E. Kimmel, Parameter estimation for a prosthetic ankle, Ann. Biomed. Eng. 23 (5) (1995) 691–696 doi: 10.1007/BF02584466.

[32] P.G. Adamczyk, S.H. Collins, A.D. Kuo, The advantages of a rolling foot in human walking, J. Exp. Biol. 209 (20) (2006) 3953.

[33] P. G. Adamczyk and A. D. Kuo, "Mechanical and energetic consequences of rolling foot shape in human walking. J. Exp. Biol. p. jeb.082347, Apr. 2013, doi: 10.1242/jeb.082347.

[34] M. Garcia, A. Chatterjee, A. Ruina, M. Coleman, and others. The simplest walking model: stability, complexity, and scaling. J. Biomech. Eng. Trans. ASME., 120 (2), 1998, 281-288.

[35] A.D. Kuo, Energetics of actively powered locomotion using the simplest walking model, J. Biomech. Eng. 124 (1) (Feb 2002) 113–120.

[36] T. McGeer. Passive dynamic walking. Int. J. Robot. Res., 9(2), 1990, 62.

[37] K.E. Zelik, T.-W.P. Huang, P.G. Adamczyk, A.D. Kuo, The role of series ankle elasticity in bipedal walking, J. Theor. Biol. 346 (2014) 75–85 doi: 10.1016/j.jtbi.2013.12.014.

[38] E. Ficanha, G. Ribeiro, L. Knop, M. Rastgaar, Estimation of the two degrees-of-freedom time-varying impedance of the human ankle, J. Med. Devices 12 (1) (2018) 011010 doi: 10.1115/1.4039011.

[39] H. Lee, N. Hogan, Time-varying ankle mechanical impedance during human loco-motion, IEEE Trans. Neural Sys. Rehab. Eng. 23 (5) (2015) 755–764 doi: 10.1109/TNSRE.2014.2346927.

[40] M.A. Rastgaar, P. Ho, H. Lee, H.I. Krebs, and N. Hogan. Stochastic estimation of multi-variable human ankle mechanical impedance. in: ASME 2009 Dynamic Systems and Control Conference, pp. 45-47, 2009, doi: 10.1115/DSCC2009-2643.

[41] International Committee of the Red Cross, Transtibial Dynamic Alignment, Physio-opedia. (2015). Available from: https://www.slideshare.net/Physiopedia/transtibial-dynamic-alignment. Accessed 11.07.18.

[42] P.G. Adamczyk, M. Roland, M.E. Hahn, Novel method to evaluate angular stiffness of prosthetic feet from linear compression tests, J. Biomech. Eng. 135 (10) (2013) 104502–1104502 doi: 10.1115/1.4025104.

[43] M.D. Geil, Energy loss and stiffness properties of dynamic elastic response prosthetic feet, J. Prosthet. Ortho. 13 (3) (2001) 70–73.

[44] G.K. Klute, J.S. Berge, A.D. Segal, Heel-region properties of prosthetic feet and shoes, J. Rehabil. Res. Dev. 41 (4) (2004) 535–546.

[45] J.F. Lehmann, R. Price, S. Boswell-Bessette, A. Dralle, K. Questad, B.J. deLateur, Com-prehensive analysis of energy storing prosthetic feet: flex foot and Seattle foot versus standard SACH foot, Arch. Phys. Med. Rehabil. 74 (11) (1993) 1225–1231.

[46] H. W. L. Van Jaarsveld, H.J. Grootenboer, J.D. Vries, and H. F. J. M. Koopman. Stiff-ness and hysteresis properties of some prosthetic feet. 1990. [Online]. Available from: http://informahealthcare.com/doi/abs/10.3109/03093649009080337. [Accessed: 31-Mar-2014].

[47] N.P. Fey, G.K. Klute, R.R. Neptune, The influence of energy storage and return foot stiffness on walking mechanics and muscle activity in below-knee amputees, Clin. Biomech. 26 (10) (2011) 1025–1032 doi: 10.1016/j.clinbiomech.2011.06.007.

[48] E. Klodd, A.H. Hansen, S. Fatone, M. Edwards, Effects of prosthetic foot forefoot flex-ibility on gait of unilateral transtibial prosthesis users, J. Rehabil. Res. Dev. 47 (9) (2010) 899–910.

[49] M. Major, M. Twiste, L. Kenney, D. Howard, The effects of prosthetic ankle stiffness on stability of gait in people with trans-tibial amputation, J. Rehabil. Res. Develop. 53 (2017) 839–852 doi: 10.1682/JRRD.2015.08.0148.

[50] M.J. Major, M. Twiste, L.P.J. Kenney, D. Howard, The effects of prosthetic ankle stiff-ness on ankle and knee kinematics, prosthetic limb loading, and net metabolic cost of trans-tibial amputee gait, Clin. Biomech. 29 (1) (2014) 98–104 doi: 10.1016/j.clinbio-mech.2013.10.012.

[51] M.J. Major, N.P. Fey, Considering passive mechanical properties and patient user motor performance in lower limb prosthesis design optimization to en-hance rehabilitation outcomes, Phys. Ther. Rev. 22 (3-4) (2017) 202–216 doi: 10.1080/10833196.2017.1346033.

[52] S.U. Raschke, et al. Biomechanical characteristics, patient preference and activity level with different prosthetic feet: a randomized double blind trial with laborato-ry and community testing, J. Biomech. 48 (1) (2015) 146–152 doi: 10.1016/j.jbio-mech.2014.10.002.

[53] J.D. Ventura, G.K. Klute, R.R. Neptune, The effects of prosthetic ankle dorsiflexion and energy return on below-knee amputee leg loading, Clin. Biomech. 26 (3) (2011) 298–303 doi: 10.1016/j.clinbiomech.2010.10.003.

[54] D.C. Morgenroth, et al. The effect of prosthetic foot push-off on mechanical loading associated with knee osteoarthritis in lower extremity amputees, Gait Posture 34 (4) (2011) 502–507 doi: 10.1016/j.gaitpost.2011.07.001.

[55] A.D. Segal, et al. The effects of a controlled energy storage and return prototype pros-thetic foot on transtibial amputee ambulation, Human Move. Sci. 31 (4) (2012) 918–931 doi: 10.1016/j.humov.2011.08.005.

[56] K.E. Zelik, et al. Systematic variation of prosthetic foot spring affects center-of-mass mechanics and metabolic cost during walking, IEEE Trans. Neural Syst. Rehabil. Eng. 19 (4) (2011) 411–419 doi: 10.1109/TNSRE.2011.2159018.

[57] FlexiForce Force Sensors, Tekscan, 02-Dec-2014. [Online]. Available from: https://www.tekscan.com/force-sensors. [Accessed: 06-Jan-2020].

[58] E. M. Glanzer and P. G. Adamczyk, Design and Validation of a Semi-Active Variable Stiffness Foot Prosthesis, IEEE Trans. Neural Sys. Rehabil. Eng., 26 (12), 2018, 2351-2359, doi: 10.1109/TNSRE.2018.2877962.

[59] B. Mariani, C. Hoskovec, S. Rochat, C. Büla, J. Penders, K. Aminian, 3D gait assessment in young and elderly subjects using foot-worn inertial sensors, J. Biomech. 43 (15) (2010) 2999–3006 doi: 10.1016/j.jbiomech.2010.07.003.

[60] L. Ojeda, J. Borenstein, Non-GPS navigation for security personnel and first responders, J. Navig. 60 (03) (2007) 391–407 doi: 10.1017/S0373463307004286.

[61] J.R. Rebula, L.V. Ojeda, P.G. Adamczyk, A.D. Kuo, Measurement of foot placement and its variability with inertial sensors, Gait Posture 38 (4) (2013) 974–980 doi: 10.1016/j.gaitpost.2013.05.012.

[62] L.V. Ojeda, P.G. Adamczyk, J.R. Rebula, L.V. Nyquist, D.M. Strasburg, N.B. Alexander, Reconstruction of body motion during self-reported losses of balance in community-dwelling older adults, Med. Eng. Phys. 64 (2019) 86–92 doi: 10.1016/j.medengphy.2018.12.008.

[63] L.V. Ojeda, J.R. Rebula, A.D. Kuo, P.G. Adamczyk, Influence of contextual task constraints on preferred stride parameters and their variabilities during human walking, Med. Eng. Phys. 37 (10) (2015) 929–936 doi: 10.1016/j.medengphy.2015.06.010.

[64] W. Wang, P.G. Adamczyk, Analyzing gait in the real world using wearable movement sensors and frequently repeated movement paths, Sensors 19 (8) (2019) 1925 doi: 10.3390/s19081925.

[65] L. Ojeda, J.R. Rebula, P.G. Adamczyk, A.D. Kuo, Mobile platform for motion capture of locomotion over long distances, J. Biomech. 46 (13) (2013) 2316–2319 doi: 10.1016/j.jbiomech.2013.06.002.

[66] M. Paulich, M. Schepers, N. Rudigkeit, G. Bellusci, Xsens MTw Awinda: miniature wireless inertial-magnetic motion tracker for highly accurate 3D kinematic applications, XSens, B.V., Technical Report MW0404P.A, 2018, doi:10.13140/rg.2.2.23576.49929.

[67] R.S. Gailey, et al. The amputee mobility predictor: an instrument to assess determinants of the lower-limb amputee's ability to ambulate, Arch. Phys. Med. Rehabil. 83 (5) (2002) 613–627 doi: 10.1053/apmr.2002.32309.

[68] D. Amtmann et al., The PLUS-M: item bank of mobility for prosthetic limb users, in: QUALITY OF LIFE RESEARCH, 2014, 23, pp. 39-40.

[69] L. E. Powell and A. M. Myers, The activities-specific balance confidence (ABC) scale. J. Geront. Series A: Biol. Sci. Med. Sci., 50 (1), 1995, pp. M28-M34.

[70] J.E. Ware, C.D. Sherbourne, The MOS 36-Item short-form health survey (SF-36): I. Conceptual framework and item selection, Medical Care 30 (6) (1992) 473–483.

[71] National Institutes of Health, PROMIS," NIH PROMIS, 2020. [Online]. Available from: http://www.nihpromis.org. [Accessed: 05-Feb-2016].

[72] F. Franchignoni, D. Orlandini, G. Ferriero, T.A. Moscato, Reliability, validity, and responsiveness of the locomotor capabilities index in adults with lower-limb amputation undergoing prosthetic training, Arch. Phys. Med. Rehabil. 85 (5) (2004) 743–748.

[73] C. Gauthier-Gagnon, M.-C. Grisé, Tools to measure outcome of people with a lower limb amputation: update on the PPA and LCI, J Prosthet. Ortho. 18 (6) (2006) pp. P61-P67.

[74] M.-C.L. Grisé, C. Gauthier-Gagnon, G.G. Martineau, Prosthetic profile of people with lower extremity amputation: conception and design of a follow-up questionnaire, Arch. Phys. Med. Rehabil. 74 (8) (1993) 862–870 doi: 10.1016/0003-9993(93)90014-2.

[75] F. Franchignoni, A. Giordano, G. Ferriero, D. Orlandini, A. Amoresano, L. Perucca, Measuring mobility in people with lower limb amputation: Rasch analysis of the mobility section of the prosthesis evaluation questionnaire, J. Rehabil. Med. 39 (2) (Mar 2007) 138–144 doi: 10.2340/16501977-0033.

[76] F. Franchignoni, M. Monticone, A. Giordano, B. Rocca, Rasch validation of the prosthetic mobility questionnaire: a new outcome measure for assessing mobility in people with lower limb amputation, J. Rehabil. Med. 47 (5) (2015) 460–465 doi: 10.2340/16501977-1954.

[77] R.S. Hanspal, K. Fisher, R. Nieveen, Prosthetic socket fit comfort score, Disabil. Rehabil. 25 (22) (2003) 1278–1280 doi: 10.1080/09638280310001603983.

[78] O.W. Holmes, The human wheel, its spokes and felloes, Atlantic Monthly (May-1863) 567–580.

[79] M.-S. Ju, J.M. Mansour, Simulation of the double limb support phase of human gait, J. Biomech. Eng. 110 (3) (1988) 223–229.

[80] A.H. Hansen, D.S. Childress, E.H. Knox, Prosthetic foot roll-over shapes with implications for alignment of trans-tibial prostheses, Prosthet Ortho. Int. 24 (3) (Dec 2000) 205–215.

[81] C. Curtze, A.L. Hof, H.G. van Keeken, J.P.K. Halbertsma, K. Postema, B. Otten, Comparative roll-over analysis of prosthetic feet, J. Biomech. 42 (11) (2009) 1746–1753 doi: 10.1016/j.jbiomech.2009.04.009.

[82] A.H. Hansen, D.S. Childress, E.H. Knox, Roll-over shapes of human locomotor systems: effects of walking speed, Clin Biomech (Bristol, Avon) 19 (4) (May 2004) 407–414 doi: 10.1016/j.clinbiomech.2003.12.001.

[83] A.H. Hansen, M.R. Meier, M. Sam, D.S. Childress, M.L. Edwards, Alignment of transtibial prostheses based on roll-over shape principles, Prosthet. Ortho. Int. 27 (2) (2003) 89–99 doi: 10.1080/03093640308726664.

[84] A.H. Hansen, D.S. Childress, Effects of adding weight to the torso on roll-over characteristics of walkin, J. Rehabil. Res. Develop. 42 (3) (2005).

[85] A.H. Hansen, D.S. Childress, Effects of shoe heel height on biologic rollover characteristics during walking, J Rehabil. Res. Develop. 41 (4) (2004) 547–554.

[86] L.L. Flynn, R. Jafari, R. Mukherjee, Active synthetic-wheel biped with torso, IEEE Trans. Robot. 26 (5) (2010) 816–826 doi: 10.1109/TRO.2010.2061272.

[87] A.H. Hansen, C.C. Wang, Effective rocker shapes used by able-bodied persons for walking and fore-aft swaying: implications for design of ankle-foot prostheses, Gait Posture 32 (2) (2010) 181–184 doi: 10.1016/j.gaitpost.2010.04.014.

[88] H. Lee, E.J. Rouse, H.I. Krebs, Summary of human ankle mechanical impedance during walking, IEEE J. Trans. Eng. Health Med. 4 (2016) 1–7 doi: 10.1109/JTEHM.2016.2601613.

[89] E. Perreault, L. Hargrove, D. Ludvig, H. Lee, and J. Sensinger, Considering Limb Impedance in the Design and Control of Prosthetic Devices, in: Neuro-Robotics, Springer, Dordrecht, 2014, pp. 59-83.

[90] A.L. Shorter, E.J. Rouse, Mechanical impedance of the ankle during the terminal stance phase of walking, IEEE Trans. Neural Sys. Rehabil. Eng. 26 (1) (2018) 135–143 doi: 10.1109/TNSRE.2017.2758325.

[91] K. Shamaei, G.S. Sawicki, A.M. Dollar, Estimation of Quasi-Stiffness and propulsive work of the human ankle in the stance phase of walking, PLoS One 8 (3) (Mar 2013) e59935 doi: 10.1371/journal.pone.0059935.

[92] D.J. Braun, S. Apte, O. Adiyatov, A. Dahiya, and N. Hogan, Compliant actuation for energy efficient impedance modulation, in 2016 IEEE International Conference on Robotics and Automation (ICRA), 2016, pp. 636-641, doi: 10.1109/ICRA.2016.7487188.

[93] C. Everarts, B. Dehez, and R. Ronsse, Variable Stiffness Actuator applied to an active ankle prosthesis: Principle, energy-efficiency, and control, in 2012 IEEE/RSJ

International Conference on Intelligent Robots and Systems, 2012, pp. 323-328, doi: 10.1109/IROS.2012.6385789.

[94] A. Jafari, N.G. Tsagarakis, and D. G. Caldwell, AwAS-II: A new Actuator with Adjustable Stiffness based on the novel principle of adaptable pivot point and variable lever ratio, in: 2011 IEEE International Conference on Robotics and Automation (ICRA), 2011, pp. 4638-4643, doi: 10.1109/ICRA.2011.5979994.

[95] J. C. Perry, Effect of Early-stance Ankle Mechanics on the Design of Variable-stiffness Below-knee Prostheses, PhD Thesis, University of Washington, 2002.

[96] R. Schiavi, G. Grioli, S. Sen, and A. Bicchi, VSA-II: a novel prototype of variable stiffness actuator for safe and performing robots interacting with humans, in: 2008 IEEE International Conference on Robotics and Automation, 2008, pp. 2171-2176, doi: 10.1109/ROBOT.2008.4543528.

[97] N.P. Fey, G.K. Klute, R.R. Neptune, Altering prosthetic foot stiffness influences foot and muscle function during below-knee amputee walking: a modeling and simulation analysis, J. Biomech. 46 (4) (2013) 637–644 doi: 10.1016/j.jbiomech.2012.11.051.

[98] D.A. Boone, et al. Influence of malalignment on socket reaction moments during gait in amputees with transtibial prostheses, Gait Posture 37 (4) (2013) 620–626 doi: 10.1016/j.gaitpost.2012.10.002.

[99] D.A. Boone, et al. Perception of socket alignment perturbations in amputees with transtibial prostheses, J. Rehabil. Res. Dev. 49 (6) (2012) 843 doi: 10.1682/JRRD.2011.08.0143.

[100] T. Kobayashi, A.K. Arabian, M.S. Orendurff, T.G. Rosenbaum-Chou, D.A. Boone, Effect of alignment changes on socket reaction moments while walking in transtibial prostheses with energy storage and return feet, Clin. Biomech. 29 (1) (2014) 47–56 doi: 10.1016/j.clinbiomech.2013.11.005.

[101] P.G. Adamczyk, A.D. Kuo, Mechanisms of gait asymmetry due to push-off deficiency in unilateral amputees, IEEE Trans. Neural Sys. Rehabil. Eng. 23 (5) (2015) 776–785.

[102] A.H. Hansen, E.A. Nickel, Development of a bimodal ankle-foot prosthesis for walking and standing/swaying, J. Med. Devices 7 (3) (2013) 035001-035001-5, doi: 10.1115/1.4024646.

[103] K2 Sensation With D/P Flexion, Össur. [Online]. Available from: https://www.ossur.com/prosthetic-solutions/products/balance-solutions/k2-sensation-with-dp-flexion. [Accessed: 02-Feb-2019].

[104] Kinterra Foot/Ankle System, Freedom Innovations, 21-Aug-2014. [Online]. Available from: https://www.freedom-innovations.com/kinterra/. [Accessed: 02-Feb-2019].

[105] MotionFoot ® MX, Fillauer. [Online]. Available from: https://fillauer.com/Lower-Extremity-Prosthetics/feet/motionfoot.html. [Accessed: 02-Feb-2019].

[106] C.E. Bauby, A.D. Kuo, Active control of lateral balance in human walking, J Biomech. 33 (11) (Nov 2000) 1433–1440.

[107] H. Reimann, T. Fettrow, E.D. Thompson, J.J. Jeka, Neural control of balance during walking, Front Physiol. 9 (Sep 2018) 1271 doi: 10.3389/fphys.2018.01271.

[108] H. Reimann, T.D. Fettrow, E.D. Thompson, P. Agada, B.J. McFadyen, J.J. Jeka, Complementary mechanisms for upright balance during walking, PLOS One 12 (2) (2017) e0172215 doi: 10.1371/journal.pone.0172215.

[109] L. Allet, H. Kim, J. Ashton-Miller, T. De Mott, J.K. Richardson, Step length after discrete perturbation predicts accidental falls and fall-related injury in elderly people with a range of peripheral neuropathy, J. Diab. Compl. 28 (1) (2014) 79–84 doi: 10.1016/j.jdiacomp.2013.09.001.

[110] H. Kim, J.A. Ashton-Miller, A shoe sole-based apparatus and method for randomly perturbing the stance phase of gait: Test-retest reliability in young adults, J. Biomech. 45 (10) (2012) 1850–1853 doi: 10.1016/j.jbiomech.2012.05.003.

[111] K. H. Yeates, A.D. Segal, R.R. Neptune, and G. K. Klute. A coronally clutching ankle to improve amputee balance on coronally uneven and unpredictable terrain, J. Med. Devices, 12(3), 2018, 031001-031001-12, doi: 10.1115/1.4040183.

[112] T. R. Clites et al. Proprioception from a neurally controlled lower-extremity prosthesis," Sci. Trans. Med., 10(443), 2018, eaap8373, doi: 10.1126/scitranslmed.aap8373.

[113] S. H. Collins, M. Kim, T. Chen, and T. Chen. An ankle-foot prosthesis emulator with control of plantarflexion and inversion-eversion torque, in: Robotics and Automation (ICRA), 2015 IEEE International Conference on, 2015, pp. 1210-1216.

[114] E.M. Ficanha, M. Rastgaar, K.R. Kaufman, A two-axis cable-driven ankle-foot mechanism, Robot. Biomim. 1 (1) (Nov 2014) 1–13 doi: 10.1186/s40638-014-0017-0.

[115] J.R. Franz, R. Kram, Advanced age and the mechanics of uphill walking: a joint-level, inverse dynamic analysis, Gait Posture 39 (1) (2014) 135–140 doi: 10.1016/j.gaitpost.2013.06.012.

[116] M. Haggerty, D.C. Dickin, J. Popp, H. Wang, The influence of incline walking on joint mechanics, Gait Posture 39 (4) (2014) 1017–1021 doi: 10.1016/j.gaitpost.2013.12.027.

[117] A.N. Lay, C.J. Hass, R.J. Gregor, The effects of sloped surfaces on locomotion: a kinematic and kinetic analysis, J. Biomech. 39 (9) (2006) 1621–1628 doi: 10.1016/j.jbiomech.2005.05.005.

[118] A. Silder, T. Besier, S.L. Delp, Predicting the metabolic cost of incline walking from muscle activity and walking mechanics, J. Biomech. 45 (10) (Jun 2012) 1842–1849 doi: 10.1016/j.jbiomech.2012.03.032.

[119] M. Damavandi, P.C. Dixon, D.J. Pearsall, Kinematic adaptations of the hindfoot, forefoot, and hallux during cross-slope walking, Gait Posture 32 (3) (2010) 411–415 doi: 10.1016/j.gaitpost.2010.07.004.

[120] D. Rosenbaum, S. Hautmann, M. Gold, L. Claes, Effects of walking speed on plantar pressure patterns and hindfoot angular motion, Gait Posture 2 (3) (1994) 191–197 doi: 10.1016/0966-6362(94)90007-8.

[121] P.G. Adamczyk, Ankle control in walking and running: speed- and gait-related changes in dynamic mean ankle moment arm (DMAMA), J. Biomech. Eng. (2019) doi: 10.1115/1.4045817.

[122] T. Kobayashi, M.S. Orendurff, M. Zhang, D.A. Boone, Effect of transtibial prosthesis alignment changes on out-of-plane socket reaction moments during walking in amputees, J. Biomech. 45 (15) (Oct 2012) 2603–2609 doi: 10.1016/j.jbiomech.2012.08.014.

[123] E.S. Neumann, J. Brink, K. Yalamanchili, J.S. Lee, Use of a load cell and force-moment analysis to examine transtibial prosthesis foot rollover kinetics for anterior-posterior alignment perturbations, JPO J. Prosthet. Ortho. 24 (4) (2012) 160–174 doi: 10.1097/JPO.0b013e31826f66f0.

[124] M. J. Greene and P. G. Adamczyk, Design and control of a semi-active two-axis prosthetic ankle, IEEE Transactions on Mechatronics, in review.

[125] S. Au, M. Berniker, H. Herr, Powered ankle-foot prosthesis to assist level-ground and stair-descent gaits, Neural Networks 21 (4) (2008) 654–666.

[126] B. E. Lawson, J. Mitchell, D. Truex, A. Shultz, E. Ledoux, and M. Goldfarb. A robotic leg prosthesis: design, control, and implementation, IEEE Robot. Automat. Mag., 21 (4), 2014, 70-81, doi: 10.1109/MRA.2014.2360303.

[127] H. A. Varol, F. Sup, and M. Goldfarb. Multiclass real-time intent recognition of a powered lower limb prosthesis, IEEE Trans. Biomed. Eng., 57(3), 2010, 542-551, doi: 10.1109/TBME.2009.2034734.

[128] J. P. Diaz, R.L. da Silva, B. Zhong, H.H. Huang, and E. Lobaton. Visual terrain identification and surface inclination estimation for improving human locomotion with a lower-limb prosthetic, in: 2018 40th Annual International Conference of the IEEE Engineering in Medicine and Biology Society (EMBC), 2018, pp. 1817-1820, doi: 10.1109/EMBC.2018.8512614.

[129] N. E. Krausz, T. Lenzi, and L. J. Hargrove. Depth sensing for improved control of lower limb prostheses. IEEE Trans. Biomed. Eng., 62(11), 2015, 2576-2587, doi: 10.1109/TBME.2015.2448457.

[130] M. Liu, D. Wang, and H. H. Huang. Development of an environment-aware locomotion mode recognition system for powered lower limb prostheses. IEEE Trans. Neural Sys., 24(4), 2016, 434-443. doi: 10.1109/TNSRE.2015.2420539.

[131] T.R. Clites, H.M. Herr, S.S. Srinivasan, A.N. Zorzos, M.J. Carty, The ewing amputation: the first human implementation of the agonist-antagonist myoneural interface, Plast. Recons. Surg.—Global Open 6 (11) (Nov 2018) e1997 doi: 10.1097/GOX.0000000000001997.

[132] S. Huang, J.P. Wensman, D.P. Ferris, An experimental powered lower limb prosthesis using proportional myoelectric control, J. Med. Devices 8 (2) (2014) 024501.

Index

Note: Page numbers followed by "f" indicate figures, "t" indicate tables.